Nanotechnology Science and Technology

Ba(Ti,Zr)O₃ - Functional Materials: From Nanopowders to Bulk Ceramics

NANOTECHNOLOGY SCIENCE AND TECHNOLOGY

Nanotechnology Research Collection - 2009/2010. DVD edition
James N. Ling (Editor)
2009. 978-1-60741-293-9

Strategic Plan for NIOSH Nanotechnology Research and Guidance
Martin W. Lang (Author)
2009. 978-1-60692-678-9

Safe Nanotechnology
Arthur J. Cornwelle (Author)
2009. 978-1-60692-662-8

Nanofibers: Fabrication, Performance, and Applications
W. N. Chang (Editor)
2009. 978-1-60741-947-1

Safe Nanotechnology in the Workplace
Nathan I. Bialor (Editor)
2009. 978-1-60692-679-6

**National Nanotechnology Initiative:
Assessment and Recommendations**
Jerrod W. Kleike (Editor)
2009. 978-1-60692-727-4

Nanotechnology in the USA: Developments, Policies and Issues
Carl H. Jennings (Editor)
2009. 978-1-60692-800-4

Nanotechnology: Environmental Health and Safety Aspects
Phillip S. Terrazas (Editor)
2009. 978-1-60692-808-0

Barrier Properties of Polymer Clay Nanocomposites
Vikas Mittal (Author)
2009. 978-1-60876-021-3

TiO$_2$ Nanocrystals: Synthesis and Enhanced Functionality
Ji-Guang Li , Xiaodong Li and Xudong Sun (Authors)
2010. 978-1-60876-838-7

Nanotechnology: Nanofabrication, Patterning and Self Assembly
Charles J. Dixon and Ollin W. Curtines (Editors)
2010. 978-1-60692-162-3

Nanostructured Materials: Classification, Properties and Fabrication
Anees A. Ansari, M. Naziruddin Khan, M. Alhoshan,
A.S. Aldwayyan and M.S. Alsalhi (Authors)
2010. 978-1-61668-763-2

Gold Nanoparticles: Properties, Characterization and Fabrication
P. E. Chow (Editor)
2010. 978-1-61668-009-1

Nanomaterials: Properties, Preparation and Processes
Vinicius Cabral and Renan Silva (Editors)
2010. 978-1-60876-627-7

Low-K Nanoporous Interdielectrics: Materials, Thin Film
Fabrications, Structures and Properties
Moonhor Ree, Jinhwan Yoon and Kyuyoung Heo (Authors)
2010. 978-1-61668-749-6

Micro Electro Mechanical Systems (MEMS): Technology, Fabrication
Processes and Applications
Britt Ekwall and Mikkel Cronquist (Editors)
2010. 978-1-60876-474-7

Nanoparticles: Properties, Classification,
Characterization, and Fabrication
Aiden E. Kestell and Gabriel T. DeLorey (Editors)
2010. 978-1-61668-344-3

**Ba(Ti,Zr)O$_3$ - Functional Materials:
From Nanopowders to Bulk Ceramics**
Adelina Ianculescu and Liliana Mitoseriu (Authors)
2010. 978-1-61668-752-6

Applications of Electrospun Nanofiber Membranes for Bioseparations
Todd J. Menkhaus, Lifeng Zhang and Hao Fong (Authors)
2010. 978-1-60876-782-3

**Superparamagnetic Iron Oxide Nanoparticles: Synthesis, Surface
Engineering, Cytotoxicity and Biomedical Applications**
*Morteza Mahmoudi, Pieter Stroeve, Abbas S. Milani
and Ali S. Arbab (Authors)*
2010. 978-1-61668-964-3

Gold Nanoparticles as an Antigen Carrier and an Adjuvant
*L. A. Dykman, S. A. Staroverov, V. A. Bogatyrev
and S. Yu. Shchyogolev (Authors)*
2010. 978-1-61668-771-7

**Energetics and Percolation Properties of
Hydrophobic Nanoporous Media**
V. N. Tronin and V. D. Borman (Authors)
2010. 978-1-61668-866-0

**Dynamics of Infiltration of a Nanoporous
Media with a Nonwetting Liquid**
V. D. Borman and V. N. Tronin (Authors)
2010. 978-1-61668-865-3

Ion-Synthesis of Silver Nanoparticles and their Optical Properties
Andrey L. Stepanov (Author)
2010. 978-1-61668-862-2

Nanoscale Signal Processing for Hybrid Computer Communications
Preecha Yupapin, Somsak Mitatha and Jalil Ali (Authors)
2010. 978-1-61728-013-9

Nanotechnology Science and Technology

Ba(Ti,Zr)O₃ - Functional Materials: From Nanopowders to Bulk Ceramics

Adelina Ianculescu

AND

Liliana Mitoseriu

Nova Science Publishers, Inc.

New York

For permission to use material from this book please contact us:
Telephone 631-231-7269; Fax 631-231-8175
Web Site: http://www.novapublishers.com

NOTICE TO THE READER

The Publisher has taken reasonable care in the preparation of this book, but makes no expressed or implied warranty of any kind and assumes no responsibility for any errors or omissions. No liability is assumed for incidental or consequential damages in connection with or arising out of information contained in this book. The Publisher shall not be liable for any special, consequential, or exemplary damages resulting, in whole or in part, from the readers' use of, or reliance upon, this material.

Independent verification should be sought for any data, advice or recommendations contained in this book. In addition, no responsibility is assumed by the publisher for any injury and/or damage to persons or property arising from any methods, products, instructions, ideas or otherwise contained in this publication.

This publication is designed to provide accurate and authoritative information with regard to the subject matter covered herein. It is sold with the clear understanding that the Publisher is not engaged in rendering legal or any other professional services. If legal or any other expert assistance is required, the services of a competent person should be sought. FROM A DECLARATION OF PARTICIPANTS JOINTLY ADOPTED BY A COMMITTEE OF THE AMERICAN BAR ASSOCIATION AND A COMMITTEE OF PUBLISHERS.

Library of Congress Cataloging-in-Publication Data

Available upon Request
ISBN: 978-1-61668-752-6

Published by Nova Science Publishers, Inc. † New York

CONTENTS

PREFACE

Ba(Ti,Zr)O_3 solid solutions were prepared by a few alternative techniques: sol-precipitation, coprecipitation via oxalate route, the modified Pechini method and the classical solid state reaction procedure. For the three types of nanopowders prepared via wet-chemical techniques, thermal analysis results suggest that in the case of both the oxalate route and modified Pechini procedure the formation kinetic of the BaTi$_{0.85}$Zr$_{0.15}$O$_3$ solid solution is more rapid, requiring a lower thermal treatment temperature, than that for the sol-precipitation synthesis. Moreover, the room temperature X-ray diffraction data also revealed that unlike the powder prepared by sol-precipitation which was di-phasic even after annealing at 1100°C, a lower annealing temperature (of 850°C) was required to obtain single-phase and well-crystallized Ba(Ti$_{0.85}$Zr$_{0.15}$)O$_3$ nanopowders resulted from the mixed molecular precursor prepared via oxalate route, as well as from the polymeric precursor synthesized by the modified Pechini method. The same cubic symmetry of the unit cell, but different morphology, agglomeration tendency and particle size of 20, 66 and 92 nm were exhibited by the powders prepared via sol-precipitation, oxalate route and the modified Pechini method, respectively. Further, Ba(Ti,Zr)O_3 ceramics with various grain sizes were prepared by solid state reaction. The influence of both zirconium content and grain size (for the fixed nominal BaTi$_{0.9}$Zr$_{0.1}$O$_3$ composition) on the microstructure and dielectric behavior was pointed out. The dielectric data showed that a ferroelectric-relaxor crossover with co-existance of both states takes place either induced by composition (by increasing Zr addition) or by reducing grain size down to nanoscale. A new type of characterization using the First Order Reversal Curves (FORC) diagrams based on recording minor hysteresis loops has been employed for the characterization of the switching process in Ba(Ti$_{1-x}$Zr$_x$)O$_3$

ceramics. This analysis proved that all the samples present a macroscopic ferroelectric behavior; but with reducing the grain size and by increasing the Zr addition, the system tends to modify its ferroelectric character as due to the increasing local heterogeneity and disorder and therefore, the system tends to its relaxor state. Microstructure and functional properties at room temperature of the ceramics prepared by the solid state reaction were compared with those obtained for the classically sintered ceramics prepared from the nanopowders synthesized by modified Pechini method. The effect of the sintering procedure on the microstructural and dielectric behavior of $BaTi_{0.85}Zr_{0.15}O_3$ ceramics prepared by modified the Pechini method was studied, by an attempt to obtain ultrafine-grained nanoceramics with grain size down to 100 nm by spark plasma sintering.

1. STATE OF THE ART
FOR THE BaTiO₃-BaZrO₃ SYSTEM

Barium titanate $BaTiO_3$ is one of the most used ferroelectrics in the microelectronic industry (production of ≈ 11000 tons in 2002), especially as dielectric material in multilayer ceramic capacitors – MLCC (figure 1), piezoelectric actuators, electroluminescent panels, pyroelectric detectors, embedded capacitance in printed circuit boards, positive temperature coefficient of resistivity (PTCR) sensors, controllers and pulse generating devices [1-6].

One of the present-day tendencies in the microelectronic industry is the miniaturization of the components and devices. Thus, the current development of the ceramic capacitors occurs in two directions: (i) development of as large capacities as possible, (ii) obtaining of components occupying a volume as small as possible. To implement the first condition, new high permittivity materials should be found. From the materials that comply with this condition, the main candidates are ferroelectrics, relaxors or ferroelectric-relaxor solid solutions in the range of concentrations corresponding to the phases overlapping (Morphotropic Phase Boundary - MPB). For a high volumetric capacity C_v, since $C_v \sim \varepsilon_r \cdot n/d^2$, where n is the number of layers and d the thickness of one ceramic layer (figure 1), the thickness of the dielectric layer should be as small as possible and the number of layers as high as possible. In the near future, the thickness of a single dielectric layer will become sub-micrometric and the number of dielectric layers will exceed 1000 [7-9]. For mechanical stability and properties reproducibility, at least 7-10 ceramic grains within one single-layer are necessary and this means that ceramics with average grain sizes below 100 nm will be employed.

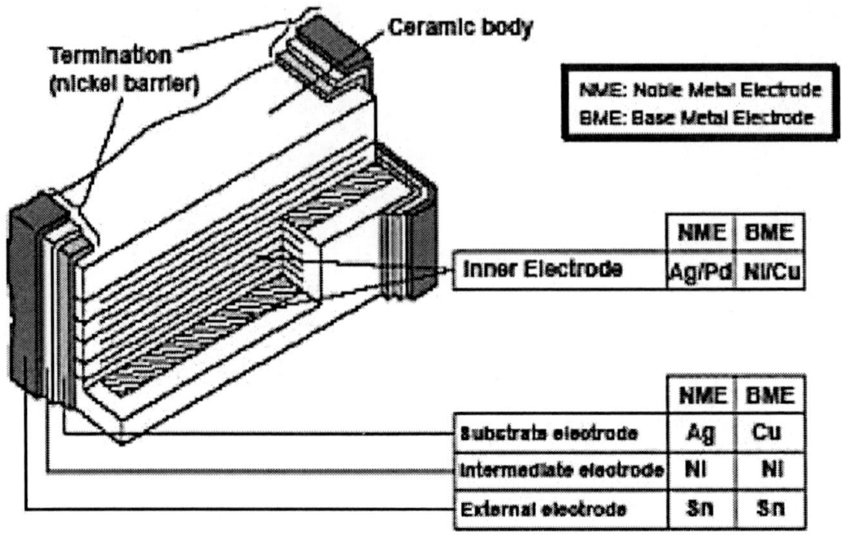

	NME	BME
Inner Electrode	Ag/Pd	Ni/Cu

	NME	BME
Substrate electrode	Ag	Cu
Intermediate electrode	Ni	Ni
External electrode	Sn	Sn

Figure 1. Cross-section scheme of a multilayer ceramic capacitor [7-9].

$BaTiO_3$ is ferroelectric at room temperature, with high dielectric and piezoelectric constants, which strongly depend on the microstructure and composition. It has a Curie temperature of ~130°C, where these constants present important thermal anomalies. The remanent polarization, the dielectric and piezoelectric constants are not as high as those reported for the ferroelectric PZT or Pb-based relaxors. Still these properties can be improved by doping with other elements or by microstructural engineering.

The type of the ferroelectric − paraelectric phase transition and the corresponding critical temperature (Curie temperature T_C) of $BaTiO_3$ can be modified via partial substitution of Ba ions (A-site doping), Ti ions (B-site doping) or both Ba and Ti ions (simultaneous A and B-site doping). A-site doping with cations of the same valence as Ba causes a shift of the Curie temperature either toward lower temperature values (T_C decrease for Sr substitution) or toward higher temperatures, stabilizing the ferroelectric state (T_C increasing for Pb substitution), without any significant broadening of the transition. With B-site doping, the cooperative dipolar long-range order formed by the off-center displacement of Ti^{4+} ions in their TiO_6 octahedra is disrupted and this often leads to a broadening of the transition in the range of Curie temperature, thus resulting the so-called *diffuse phase transition*. Therefore, by the formation of B-site homovalent solid solutions via Zr^{4+} [10-14], Sn^{4+} [14-20], Hf^{4+} [21-25] and Ce^{4+} [26-28] incorporation in the $BaTiO_3$

perovskite lattice, one can obtain a composition-induced ferroelectric-relaxor crossover. In the relaxor state the system exhibits almost zero remanent polarization (and no hysteresis in the P(E) and ε(E) dependences) and very high dielectric, piezoelectric and pyroelectric constants, mainly for the compositions placed at Morphotropic Phase Boundary (MPB). The relaxor state is also characterized by a range for the ferroelectric-paraelectric phase transition (around the temperature corresponding to the maximum permittivity T_m) instead of the Curie temperature.

The shift of the Curie temperature by various substitutions, the superior properties at MPB and the lack of any hysteretic behavior in the relaxor state offer excellent freedom degree of material engineering in order to bring at room temperature a desired state (relaxor, ferroelectric, paraelectric state, MPB) or material properties for specific applications in electronic industry.

In spite BaTiO$_3$ is a classical well studied ferroelectric, the properties of its solid solutions and their dependence on the preparation method and microstructure, particularly at submicron range, are topics of high novelty, interest and actuality. The optimisation of the preparation parameters leads to the developing of new, environmental friendly materials, exhibiting superior functional properties, that might represent viable alternative to replace Pb-based oxides compounds in various electroceramics required by the microelectronics industry. The ferroelectric and relaxor properties are even less studied in nanosized BaTi$_{1-x}$M$_x$O$_3$ (M = Zr, Sn, Hf, Ce) materials, as thin films and nanoceramics.

BaTi$_{1-x}$Zr$_x$O$_3$ (BTZ), is one of the most important composition used as dielectric material in MLCCs and as a storage capacitor for the next DRAM generation. Complete solubility of BaZrO$_3$ in BaTiO$_3$ with the formation of continuously isomorphous Ba(Ti$_{1-x}$Zr$_x$)O$_3$ solid solution (which gradually changes its structural symmetry depending on the Ti/Zr ratio) occurs after sintering at 1400°C for 10 hours, as reported by early studies of phase equilibrium in the ternary BaO-TiO$_2$-ZrO$_2$ system [29, 30].

It is well known that Zr is an effective solute in BaTiO$_3$ not only to shift the Curie temperature below room temperature [31], but also to significantly broaden the permittivity-temperature dependence. Thereby, the material can be adapted to meet the Z5U specification, resulting in widening the temperature stability of devices. Moreover, Zr^{4+} ion is chemically more stable than Ti^{4+} and has a larger ionic radius (0.87 Å) than that of Ti^{4+} (0.68 Å), inducing the expanding of the perovskite lattice. Therefore, the substitution of Ti by Zr would depress the conduction by small polaron hopping between Ti^{4+} and Ti^{3+} and it would also decrease the leakage current of the BaTiO$_3$ system. Zr/Ti ratio is a very important parameter in the BTZ system which tailors the ferroelectric – paraelectric phase transition type and its characteristic temperature. For lower zirconium concentration ($0 < x \leq 0.10$), $BaTi_{1-x}Zr_xO_3$ behaves like a classical ferroelectric material. Brajer and Kulscar showed that, as the zirconium content increases, the orthorhombic-tetragonal phase transition temperature increases and the tetragonal-cubic phase transition temperature concurrently decreases [32, 33]. BTZ exhibits a pinched phase transition at $x = 0.15$, where all the three dielectric constant peaks coalesce into a single broad maximum [34, 35]. Further increase of Zr content ($0.10 < x < 0.25$) would result in the decrease of the Curie temperature (which is shifted below the room temperature, for $x \geq 0.23$), as well as in an increase of the diffuseness degree of the permittivity-temperature dependence, indicating a mixed ferroelectric-relaxor behavior [34]. The ferroelectric-relaxor crossover composition of BTZ has been claimed to be placed at $x = 0.25$ [36, 37]. Large, diffuse and frequency-dispersive maxima in the temperature dependence of the dielectric permittivity indicating a relaxor-like behavior were found at higher Zr contents ($0.25 < x \leq 0.40$) [10]. In this case, the temperature T_m, corresponding to the large permittivity maximum is shifted far below the room temperature. Figure 2 shows the temperature evolution of the real and imaginary part of the dielectric constant for the relaxor $BaTi_{0.65}Zr_{0.35}O_3$ reported by *Simon et al.* [36]. Thus, for a frequency of 10 kHz, the temperature of the maximum of the dielectric permittivity is close to 200 K (figure 2).

For a ceramic sample prepared by solid state reaction method, implying classical sintering at 1450°C/6 hours, it was estimated that the phase transition occurs at room temperature, as the zirconium content reaches a value of $x \sim 0.225$ [34].

Briefly, taking into account several literature data [10, 34-37], the composition dependence of the phase transition temperature, T_C (or T_m in the case of the relaxor state) for $BaTi_{1-x}Zr_xO_3$ solid solutions was synthetically represented in figure 3.

One can notice that for $x > 0.25$, the system seems to reach its paraelectric state at room temperature. However, it is worthy to mention that the morphotropic limits of the phase transitions are very sensitive to the preparation method, sintering parameters, purity and homogeneity degree at local scale, all these parameters inducing particular structural and/or microstructural features [38].

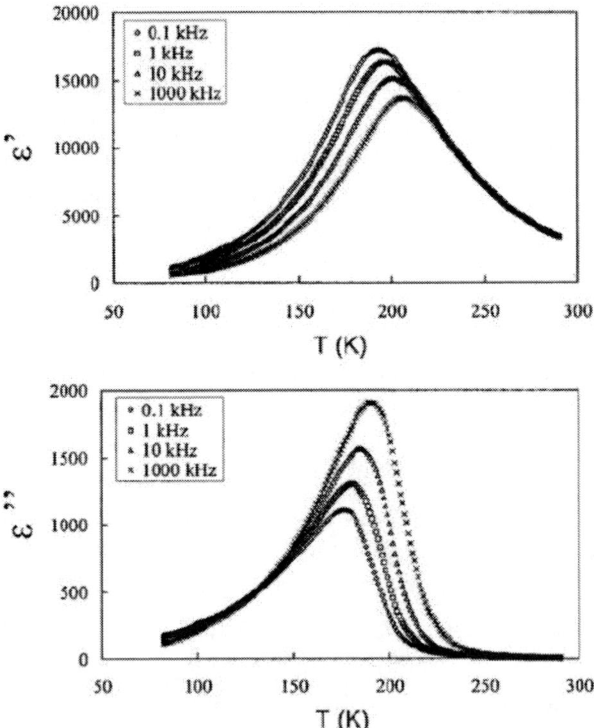

Figure 2. Evolution of the real part (a) and imaginary part (b) of the dielectric constant *versus* temperature for various frequencies, in BaTi$_{0.65}$Zr$_{0.35}$O$_3$ ceramic (*J. Phys. Condens. Matter.*, 16 (2004) 963-970 [36]).

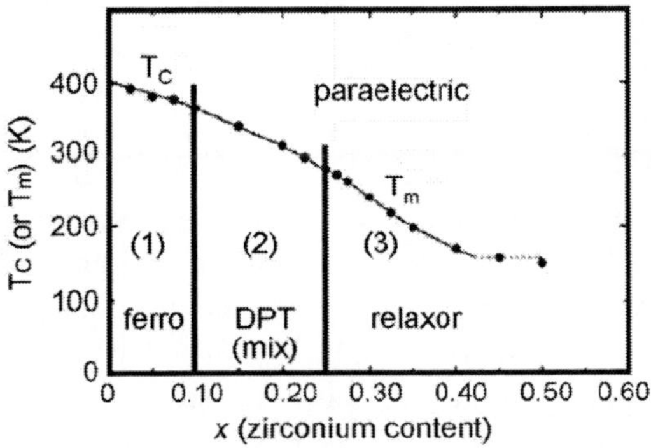

Figure 3. Evolution of the temperature corresponding to the permittivity maximum *versus* the zirconium content in BaTi$_{1-x}$Zr$_x$O$_3$ system (DPT = diffuse phase transition).

Recently, BTZ has also shown great potential for the application in tunable microwave devices [37-42], due to its large dielectric nonlinearity (*i.e.* the dependence of permittivity on dc electric field) and low dielectric loss. Thus, for $BaTi_{0.8}Zr_{0.2}O_3$ ceramic measured at room temperature under the biasing field of 20 kV/cm, high tunability (of ~82%) and low dielectric losses (tan δ = 0.0034) were reported [43]. However, unlike the canonical relaxors (such as PMN), an electric field does not seem to induce a long-range polar order in BTZ relaxors [44]. Indeed, recent studies considered the behavior of these BTZ compositions rather as corresponding to a quasi-ferroelectric than to a typical relaxor. Some evidences sustaining this assertion are: (*i*) normal ferroelectric behavior under high pressure conditions pointed out for the $BaTi_{0.65}Zr_{0.35}O_3$ ceramic by the high-pressure Raman scattering investigations [45]; (*ii*) the lack of heat capacity anomaly around T_m indicating that freezing is absent [46]; (*iii*) different dielectric dispersion behavior and a dielectric peak described rather by a Lorenz-type law, instead of the well-known modified Curie-Weiss law characteristic for the canonical relaxors [47]. Thus, in the perovskite $Ba(Ti_{0.675}Zr_{0.325})O_3$ (BTZ325) solid solution which was shown to be relaxor with a cubic structure (space group $Pm\overline{3}m$) down to low temperatures (similar to canonical relaxors) [35, 36, 44, 48], the dielectric dipolar dynamics was, however, found not subject to any critical slowing-down and glassy freezing. This implies that in contrast to other known relaxors, the low-temperature state in BTZ325 solid solution is neither dipolar glass nor ferroelectric (FE) type. Nevertheless, such a state is non-ergodic, and it can be considered as a quasi-ferroelectric state. This odd behavior was explained by Bokov *et al.* [49] in terms of nanoscale inhomogeneity of structure, which consists of a mixture of randomly distributed static and dynamic polar nanoregions separated by nonpolar Zr-rich nanoregions.

In the last decade, the properties of the $BaTi_{1-x}Zr_xO_3$ system were intensively study also in single crystals [50-55] and as thin films [56-61]. Concerning the bulk ceramics, the interest was focused mainly to the investigation of more fundamental aspects, as mechanism formation of the perovskite phase [34, 62-64] and influence of various types of defects (like oxygen vacancies or others), grain size effects or grain boundary phenomena on the switching characteristics [11-13, 65]. By analogy with magnetic systems, it was postulated the possibility to disrupt the long-range order and to induce a superparaelectric state characteristic of relaxor or spin-glass systems by reducing grain size in ferroelectric states [13]. Therefore, one can conclude that the interplay between the composition optimization and choice of the suitable processing route and parameters inducing particular microstructures should be essential factors for tailoring the dielectric behaviour.

The synthesis of Ba(Ti,Zr)O$_3$ solid solutions by solid state reaction is governed by high temperature diffusional phase formation kinetics. The fabrication of BTZ ceramics by this conventional technique can be achieved using two dissimilar procedures: (*i*) a single high temperature-processing step to obtain BaTi$_{1-x}$Zr$_x$O$_3$ mixed crystals, implying the direct, global reaction between BaCO$_3$, TiO$_2$ and ZrO$_2$ [62, 63] and (*ii*) a multi-step procedure, which involves the preliminary synthesis of BaTiO$_3$ and BaZrO$_3$ as independent products, followed by their high temperature solid state reaction after mixing appropriate amounts of the two components, in order to form the (1-*x*)BaTiO$_3$-*x*BaZrO$_3$ solid solution [66]. For the first case a relative micro-scale chemical homogeneity was reported [62, 63], whereas in the case of the second procedure, slower diffusion processes occurring because of the dissimilar preformed crystalline lattices specific to the two components lead to an increasing chemical non-homogeneity of the related ceramic [66]. Moreover, the experimental data of Neirman [66] for the (1-*x*)BaTiO$_3$ − *x*BaZrO$_3$ (*x* = 0.10 and 0.20) solid solutions pointed out that the variation of the Curie temperature with the firing temperature seems to originate in this difference in the local chemical and microstructural homogeneity.

S. Gopalan and A.V. Virkar [63] examined the interdiffusion in doped BT-BZ sintered couples and a Kirkendall porosity formation was found, indicating that transport occurs on all sub-lattices in the system.

Another approach to obtain ceramics with desired properties by conventional ceramic method refers to the microstructure control by establishing suitable sintering strategies. This task becomes difficult since the as-prepared powders are normally coarse, with inhomogeneous composition, highly agglomerated and require extensive milling. Consequently, they are not adequate for preparing dense ceramics, with homogeneous composition and tailored grain size down to submicron values. In order to overcome this disadvantage and to obtain very fine non-agglomerated powders of high purity, with controlled stoichiometry and narrow particle size distribution, several innovative methods have been proposed for the synthesis of Ba(Ti$_{1-x}$Zr$_x$)O$_3$ solid solutions [67-76].

2. PREPARATION AND CHARACTERIZATION OF BA(TI,ZR)O$_3$ NANOPOWDERS: INFLUENCE OF THE PROCESSING METHOD ON THE STRUCTURAL AND MORPHOLOGICAL PROPERTIES

2.1. SYNTHESIS

In spite of the importance and actuality of the topic, the number of papers describing the synthesis and the characteristics of BTZ nanopowders prepared by wet-chemical routes remains relatively low comparing with that of papers which reported data regarding pure BaTiO$_3$ powders. In some of these non-conventional routes as co-precipitation, sol-gel and hydrothermal method, a carefully optimization of the synthesis parameters and relatively high temperatures are required in order to avoid undesired features as stoichiometry deviations and formation of di-phase solid solutions.

In order to have a conclusive picture regarding the influence of the synthesis parameters on the characteristics of Ba(Ti,Zr)O$_3$ solid solutions, a comparative study of the mechanism formation, structure and morphology of these powders synthesized by various wet chemical routes is required. For this purpose, three methods (*i.e.* the sol-precipitation, co-precipitation via-oxalate route and the Pechini modified procedure) have been taken into account for the preparation of BTZ solid solution. In the following, BaTi$_{0.85}$Zr$_{0.15}$O$_3$ was chosen for a comparative discussion, since this composition is around the reported pinched phase transition value [34, 35]. The resulted powders are denoted as BTZ1 (sol-precipitation), BTZ2 (co-precipitation via-oxalate route) and BTZ3 (Pechini modified procedure), respectively. In all the three cases the titanium (IV) isopropoxide (Ti(i-OC$_3$H$_7$)$_4$) and zirconium propoxide (Zr(OC$_3$H$_7$)$_4$) were used as Ti and Zr source. The barium source varied as a function of the particularity of such of these methods: barium nitrate for the sol-precipitation, barium chloride for the co-precipitation via oxalate and barium carbonate for the modified Pechini procedure. All the chemicals were purchased from Merck and were GR grade ($\geq$ 99%)

BTZ1: The sol-precipitation synthesis of BaTi$_{0.85}$Zr$_{0.15}$O$_3$ powder is similar to that one reported for pure BaTiO$_3$ in ref. [77, 78] and more details of this procedure were presented elsewhere [79] (figure 4).

BTZ2: In the oxalate route, oxalic acid and isopropanol were used as chelating agent and solvent, respectively [80]. The molar ratio of the raw materials was: Ti(i-OC$_3$H$_7$)$_4$: Zr(OC$_3$H$_7$)$_4$: BaCl$_2$·2H$_2$O : H$_2$C$_2$O$_4$·2H$_2$O = 0.85 : 0.15 : 1 : 2. Figure 5 shows the synthesis steps of BaTi$_{0.85}$Zr$_{0.15}$O$_3$ nanopowder via oxalate route.

BTZ3: Despite the disadvantage of the relatively long preparation time, the polymeric precursor method (PPM) mainly based on the Pechini-type process is one of the most widely used for the preparation of pure mixed-oxides nanopowders, due to the low costs of precursors, low synthesis temperature and ionic homogeneity at molecular level. The modified Pechini method [81], usually used for oxide powders synthesis, consists in the formation of some complexes of alkaline, earth-alkaline and even transitional metals with organic bi- or tridentate chelating agents as the citric acid (CA). In the reaction mixture a polyalcohol as ethylene glycol (EG) is added in order to form bonds by polyesterification between the chelate compounds and thus causing the mixture gelation. After the gel drying a thermal treatment is performed to initiate the organic compounds pyrolysis and finally to obtain submicron oxide powders. A lot of papers refer to the synthesis by the modified Pechini method of pure BaTiO$_3$ [72-92], but only one to the best of our knowledge describes the formation of the Zr-doped BaTiO$_3$, *i.e.* BaTi$_{0.8}$Zr$_{0.2}$O$_3$, by this technique [93]. In the present chapter, for the synthesis of BaTi$_{0.85}$Zr$_{0.15}$O$_3$ nanopowders, the procedure reported earlier for the preparation of (Ba$_{1-x}$Sr$_x$)TiO$_3$ solid solutions was adopted [94-96]. For a molar precursor ratio BaCO$_3$:Ti(i-OC$_3$H$_7$)$_4$: Zr(OC$_3$H$_7$)$_4$: C$_6$H$_8$O$_7$: C$_2$H$_4$O$_2$ = 1 : 0.85 : 0.15 : 7.5 : 21 [84] the synthesis sequence is presented in figure 6. Single phase Ba(Ti$_{0.85}$Zr$_{0.15}$)O$_3$ powders were obtained after annealing at 850°C for 2h.

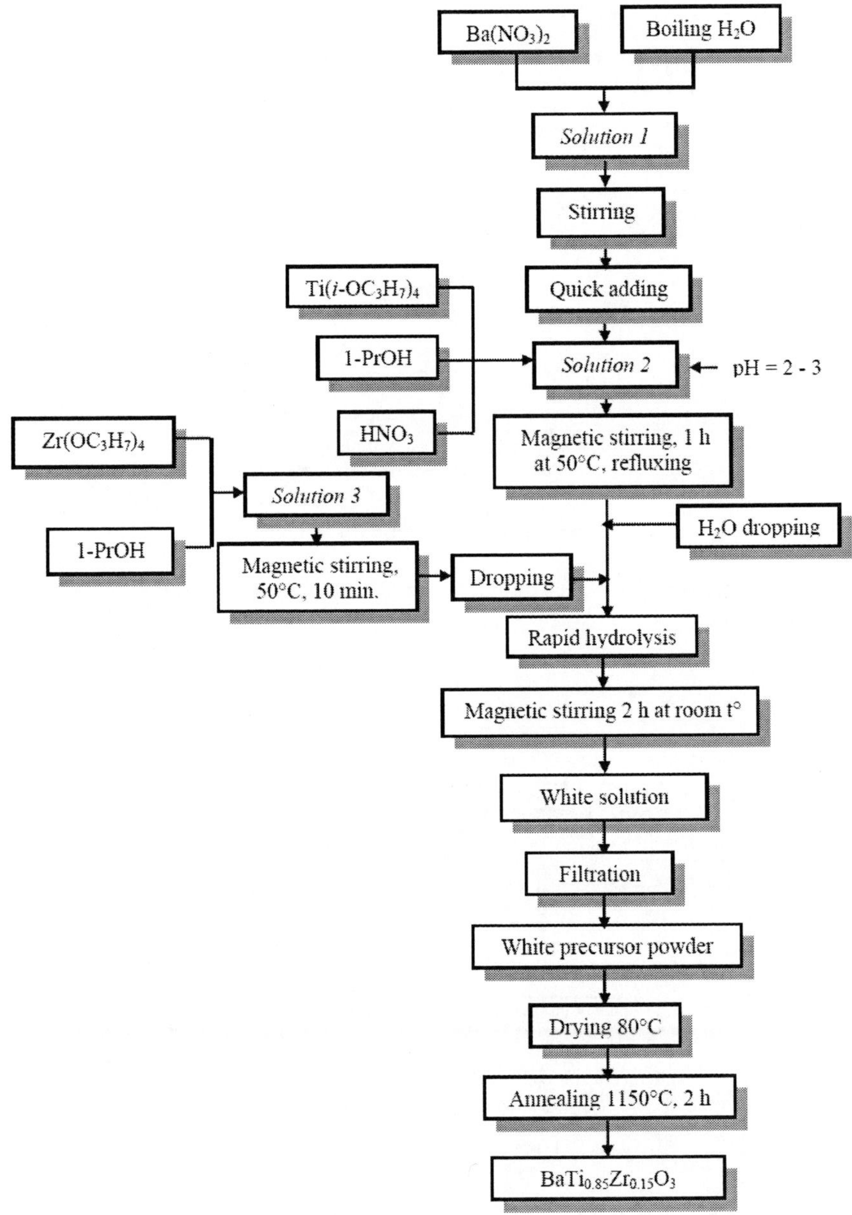

Figure 4. Flowchart for synthesis of $BaTi_{0.85}Zr_{0.15}O_3$ nanopowder by sol-precipitation method [79].

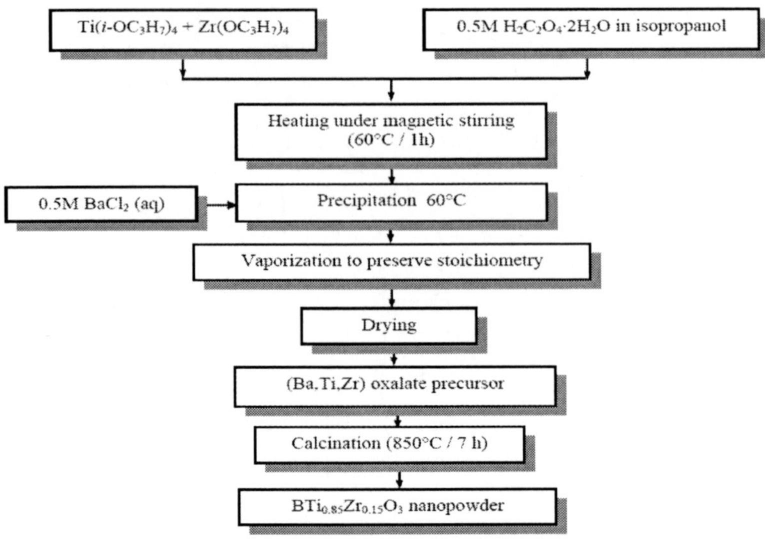

Figure 5. Flowchart for synthesis of BaTi$_{0.85}$Zr$_{0.15}$O$_3$ nanopowder by coprecipitation *via* oxalate route [80].

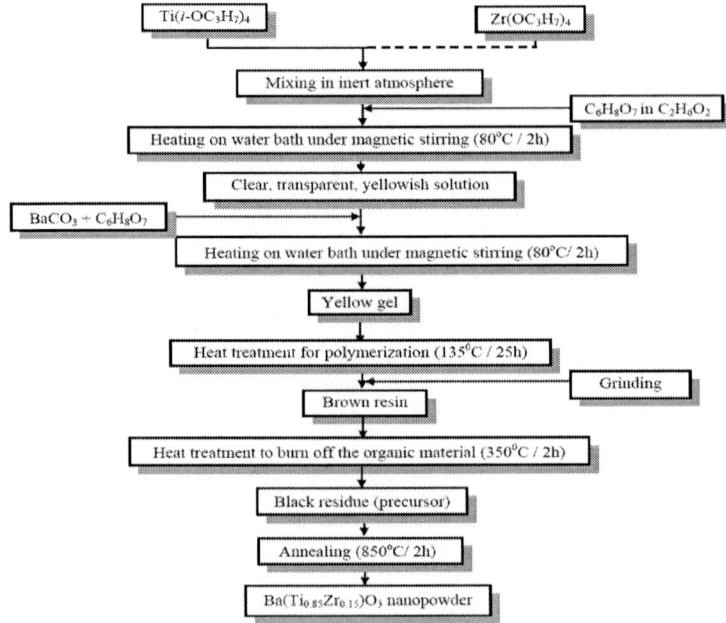

Figure 6. Flowchart for synthesis of BaTi$_{0.85}$Zr$_{0.15}$O$_3$ nanopowders by the modified Pechini method.

2.2. FORMATION MECHANISM

The gel precursor obtained by sol-precipitation exothermally decomposes in two successive steps in the temperature range of 600 – 700°C. These processes are accompanied by a mass loss of ~ 25% clearly emphasized on the TG curve (figure 7).

We assumed that after these processes the perovskite skeleton is already formed and coexists together with some residual intermediates. In the temperature range of 900 – 1050°C, the residual phases are consumed by chemical reaction, so that the formation of the perovskite BTZ1 phase is completed. This process is pointed out on the DSC curve by the presence of a small peak with the maximum located at 915.5°C and is accompanied by a small mass loss of ~ 3% determined by the CO_2 release.

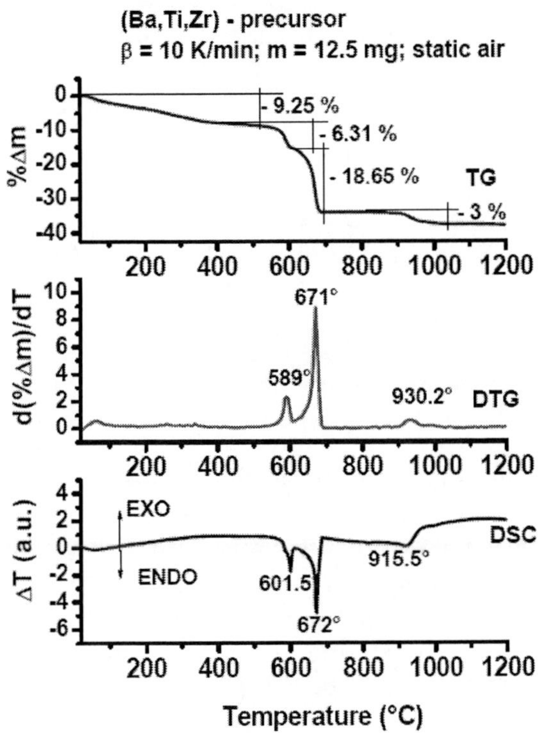

Figure 7. Thermal behavior of the sol-precipitated precursor.

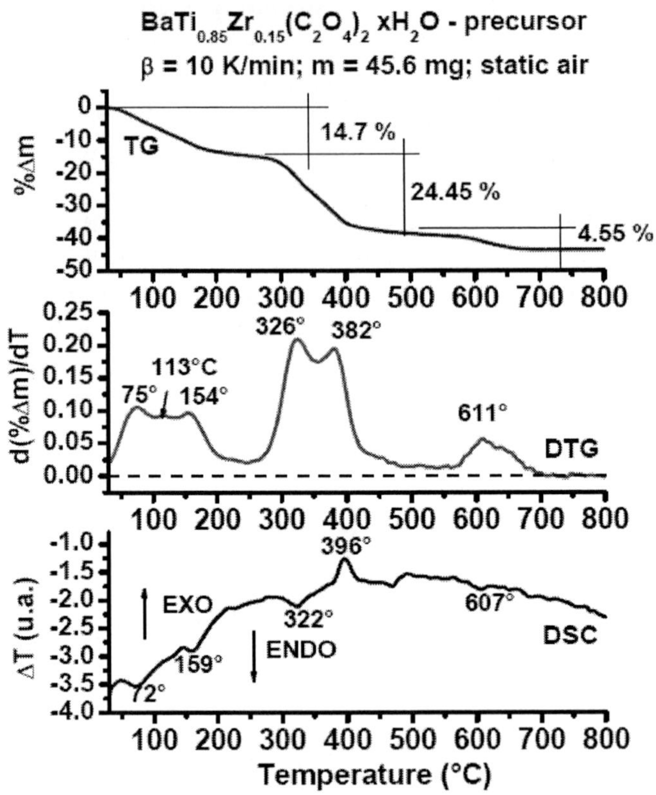

Figure 8. Thermal behavior of the mixed (Ba,Ti,Zr) oxalate precursor.

The thermal analysis curves corresponding to the amorphous precursor obtained by oxalate route emphasized three different decomposition stages (figure 8).

In the first stage (70 – 160°C) the release of both the adsorbed water (of the surface and inside the pores) and the solvent traces takes place. These processes correspond to the endothermic peaks located at 72°C and 154°C on the DSC curve. The second stage (300 – 420°C) is accompanied by the major mass loss (of ~ 25%), visible on the TG curve and was assigned tot the thermo-combustion of the organic matter. This stage occurs in two steps: the initial one occurring at 322°C is associated with the endothermic process of the carbonyl bonds breakdown, whereas the second one represents the exothermic oxidation of the organic compounds (at 396°C). In this case, the mass loss is due to the CO and CO_2 release. The third stage (575 – 700°C) is associated with the endothermic decomposition of a metastable (Ba,Ti,Zr) oxocarbonate intermediate phase, resulted after the combustion process. Consequently, the BTZ2 solid solution was formed and a small mass loss (of ~ 5%) due to the CO_2 release was recorded on the TG curve (figure 8).

For the third route, the solution resulted by mixing the two citrate solutions of (Ti,Zr) and Ba respectively, transforms by heating (80°C) into a gel, due to the chelation and esterification reactions. After polymerization (135°C), the gel converts in a dark-brown, glassy resin, as follows:

$$\text{sequence of gel after esterification} \xrightarrow{\text{Polymerization}} \text{Resin} \tag{1}$$

According to the literature data [82], during the heat treatment at 350°C/2h carried out to burn off the organic matter, the resin decomposes in several steps and the formation of the so-called "precursor" takes place.

$$
Ba^{2+}\;(Ti^{4+}+Zr^{4+})\begin{bmatrix}CH_2\!-\!COO^-\\[2pt]COH\!-\!COO^-\\[2pt]CH_2\!-\!COOH\end{bmatrix}_3 \xrightarrow[-3H_2O]{210°C} Ba^{2+}\;(Ti^{4+}+Zr^{4+})\begin{bmatrix}CH_2\!-\!COO^-\\[2pt]C\!-\!COO^-\\[2pt]CH\!-\!COOH\end{bmatrix}_3 \xrightarrow[-3CO_2]{230°C}
$$

Citrate   Aconitate

$$
Ba^{2+}\;(Ti^{4+}+Zr^{4+})\begin{bmatrix}CH_2\!-\!COO^-\\[2pt]C\!-\!COO^-\\[2pt]CH_2\end{bmatrix}_3 \xrightarrow{250-360°C} 2\;\begin{matrix}CH_2CO\\C\!-\!CO\\CH_2\end{matrix}\!\!\diagup\!\!O + (Ti,Zr)O_2 + \begin{bmatrix}CH_2\!-\!COO^-\\[2pt]C\!-\!COO^-\\[2pt]CH_2\end{bmatrix}Ba^{2+}
$$

Itaconate   Itaconic anhydride   (2)

The thermal analysis of this black precursor pointed out four thermal effects (figure 9). The first low-temperature endothermic effect is very slight and broad and it is assigned to the release of the adsorbed water. The following two exothermic effects correspond to the subsequent combustion of the organic matter and are accompanied by mass loss recorded on the TG curve. The first of them, more marked shows a maximum at 426°C, while the second with the maximum at 478°C is pointed out only as a shoulder on the DSC curve and is also highlighted by the slight slope change on the TG curve.

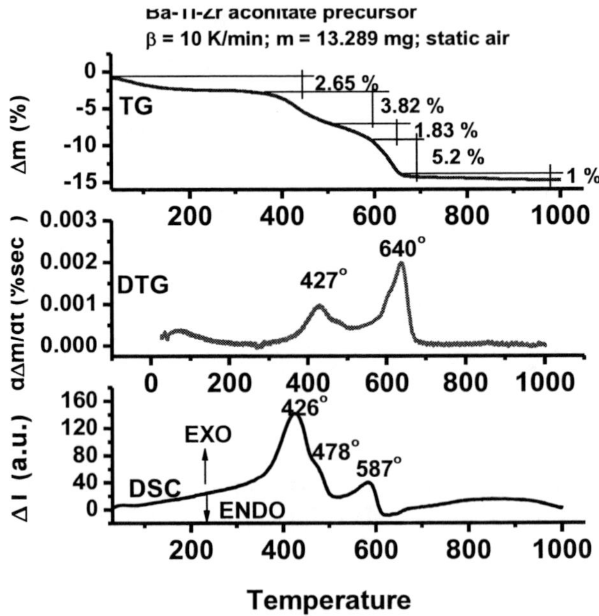

Figure 9. Thermal behavior of the mixed (Ba,Ti,Zr) polymeric precursor (black residue) resulted after calcinations at 350°C for 2 hours.

It is likely that the intermediate product which forms after the exothermic combustion consists of the metastable mixed (Ba,Ti,Zr) oxocarbonate phase [84, 91], according to the reaction (3):

$$\text{“Precursor”} + O_2 \rightarrow Ba_2(Ti_{1-x}Zr_x)_2O_5 \cdot CO_3 + xH_2O + y\,CO_2 \uparrow \quad (3)$$

This assumption is sustained by the low value of the mass loss (only of 5.2%) recorded during the last endothermic decomposition process (at ~ 640°C), which excludes the formation of the perovskite Ba(Ti,Zr)O$_3$ phase by solid state reaction (characterized by a much higher mass loss, of ~ 22 %) between intermediates as BaCO$_3$ and (Ti,Zr)O$_2$. Therefore, in this last stage, which ends at ~ 850°C, the mixed oxocarbonate intermediate decomposes into BaTiO$_3$, with CO$_2$ release, similar with the formation mechanism specific to the oxalate route (eq. 4):

$$Ba_2(Ti_{1-x}Zr_x)_2O_5 \cdot CO_3 \rightarrow 2BaTi_{1-x}Zr_xO_3 + CO_2\uparrow \qquad (4)$$

From thermal analysis data, one can conclude that in the case of both the oxalate route and modified Pechini procedure the formation kinetic of the BaTi$_{0.85}$Zr$_{0.15}$O$_3$ solid solution is more rapid, requiring a lower thermal treatment temperature, than that for the sol-precipitation synthesis.

2.3. PHASE COMPOSITION AND STRUCTURE

The X-ray diffraction patterns of Ba(Ti$_{0.85}$Zr$_{0.15}$)O$_3$ powders prepared by the mentioned methods are presented in figure 10.

The BTZ1 powder synthesized by sol-precipitation and annealed at 1100°C for 2 hours is di-phasic and shows a lower crystallinity (figure 10 (a)). Beside the major Ba(Ti,Zr)O$_3$ phase, a small amount of a secondary BaZrO$_3$ (BZ) phase identified in the XRD pattern by the presence of small, broad shoulders located at the left side (at lower 2θ values) of the BTZ main diffraction peaks, was also detected, showing that, even at this relative high temperature, the chemical reaction was not completed. The presence of this secondary BZ phase indicates that the major phase of BaTi$_{1-x}$Zr$_x$O$_3$ solid solution, is poorer in zirconium ($x < 0.15$) than the nominal composition.

In good agreement with the thermal analysis results, the room temperature X-ray diffraction data also revealed that in comparison with the BTZ1 sample, a lower annealing temperature (of 850°C) was necessary to obtain single-phase and well-crystallized Ba(Ti$_{0.85}$Zr$_{0.15}$)O$_3$ nanopowders from the mixed molecular precursor prepared via oxalate route (figure 10 (b)), as well as from the polymeric precursor synthesized by the modified Pechini method (figure 10 (c)). For the BTZ2 and BTZ3 powders the higher crystallinity was pointed out by the sharpness of the main diffraction peaks. This assertion is sustained also by the values of the crystallite average size, which indicate the highest crystallinity for the powder prepared by modified Pechini method (BTZ3).

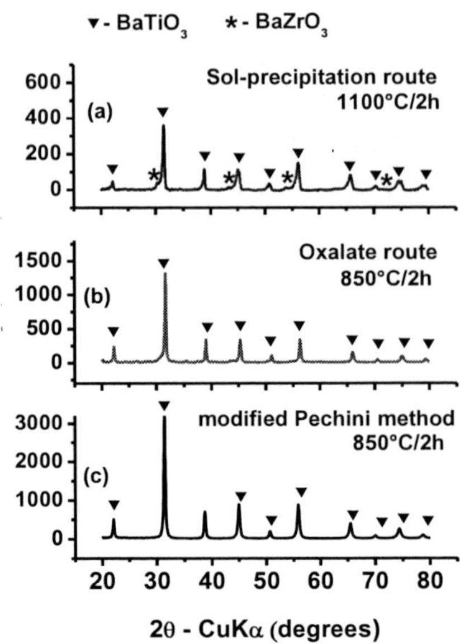

Figure 10. Room-temperature XRD patterns for the BTZ powders prepared by alternative methods.

Unlike some results reported in the literature which revealed a tetragonal structure and consequently a ferroelectric state at room temperature for the $Ba(Ti_{1-x}Zr_x)O_3$ $(0 \leq x \leq 0.15)$ solid solutions [35], the XRD patterns for all $BaTi_{0.85}Zr_{0.15}O_3$ powders investigated in this chapter show no splitting of the (200) peak in the vicinity of 2θ-CuKα ~ 45°, proving a cubic structure, *i.e.* a paraelectric state (figure 10). Indeed, the calculation of the structural parameters from the XRD data also indicates a cubic symmetry of the unit cell of the $BaTi_{0.85}Zr_{0.15}O_3$ powders, irrespective of the preparation method. These results are in good agreement with those of Cheng *et al.* who also reported a cubic structure for their nanopowders of the same composition but obtained by an aqueous coprecipitation method [72]. The disappearance of ferroelectricity and the stabilization of the paraelectric, or more exact of the superparaelectric phase at nanoscale was intensively studied mainly for $BaTiO_3$ powders. The stabilization mechanisms of the so-called "pseudocubic" phase are very complex, including surface depolarization effects, the absence of long-range cooperative interactions, elastic constrains and lattice defects as hydroxyl ions [97].

The diffraction qualitative and structural data are summarized in Table 1.

Table 1. Phase composition and structural characteristics of the BTZ powders

Sample	Phase composition	Structure	a [Å]	V [Å]	$<D>$ [Å]	$<d_{TEM}>$ [nm]
BTZ1	$Ba(Ti_{1-x}Zr_x)O_3$ +	cubic	4.0166(5)	64.80(5)	198(40)	92
	$BaZrO_3$	cubic	4.1746(8)	72.75(81)	186(43)	81
BTZ2	$BaTi_{0.85}Zr_{0.15}O_3$	cubic	3.9994(2)	63.97(2)	221(37)	20
BTZ3	$BaTi_{0.85}Zr_{0.15}O_3$	cubic	4.0293(3)	65.42(1)	668(130)	66

2.4. MORPHOLOGY

The BTZ powders prepared by the three wet chemical methods exhibit completely different morphologies, as it will be described in the following.

The SEM image of the BTZ1 sample shows larger, apparently monolithic, prismatic, cubic or lamellar-shaped micro-scaled blocks, with well defined edges and faces, coexisting with smaller, submicron aggregates (figure 11 (a)). Higher magnifications of the scanning electron microscope were not able to indicate the further structuring of these forms. Unlike this case, the SEM image of BTZ2 powder indicates the formation of more uniform (as shape and size), obviously grain-structured, micro-scaled (~ 300 nm) aggregates (figure 11 (b)). For the morphological characterization of the BTZ3 powder, a higher performance scanning electron microscope with field emission gun was used. The SEM-FEG image of the mentioned sample indicates the presence of fine submicron particles (figure 11 (c)). Higher magnification (figure 11 (c) - inset) emphasizes the sintering tendency of these particles, which is normally expected in a polymeric precursor synthesis, as the modified Pechini procedure. Due to this phenomenon, the particles seem three-dimensionally cross-linked by means of necks and bound bridges. For this reason, it is difficult to estimate an average particle size, so that further TEM investigations are required in order to have an idea if these "particles" are the ultimate structural units.

The TEM analyses revealed in the powder synthesized by the sol-precipitation method, the obtaining of non-uniform particles, with an average size of ~ 92 nm and showing a pronounced tendency to form larger agglomerates or aggregates (figure 12 (a)). A comparison between the particle mean size (estimated from the SEM analysis) and the crystallite average size (calculated from the XRD data) suggests the polycrystalline nature of the particles (each particle seems to contain ~ 4 - 5 crystallites). Lower particle size was noticed for the powders prepared by oxalate route and modified Pechini method, respectively.

Thus, BTZ2 powder consists of small, isometric (almost spherical and isolated) particles, uniform as shape and size, with a mean size of only ~ 20 nm (figure 12 (b)). Particles of intermediate mean size (of ~ 66 nm) ranged between BTZ1 and BTZ2 particle size, but more faceted, more inhomogeneous as shape and size and with higher aggregation tendency (by partial sintering) than those of BTZ2 were pointed out by the TEM image of the BTZ3 powder. The BTZ3 particles exhibit rather a polyhedral morphology, with well defined, sharp-angle joined edges which determine their blunt aspect.

Figure 11. Images of the Ba(Ti,Zr)O₃ powders synthesized by alternative routes: (a) SEM: sol-precipitation route (bar = 5 µm); (b) SEM: oxalate route (bar = 5 µm) (c) SEM-FEG general view: modified Pechini method (bar = 1 µm) and (d) SEM-FEG detail: modified Pechini method (bar = 500 nm).

Anyway, unlike BTZ1, for BTZ2 and BTZ3 powders the particle mean size seems to fit well enough the crystallite average size determined by XRD, proving the single crystal nature of these particles (Table 1).

Unlike the high crystallinity degree of the BTZ3 powder pointed out by the corresponding HRTEM image (figure 12 (d)), the spots frequency and distribution of the SAED pattern of the BTZ1 sample (figure 12 (a)) indicated that, despite the larger particles, the powder prepared by sol-precipitation route exhibits lower crystallinity. These observations are in good agreement with the XRD results presented in figure 10.

From these corroborated results, it is obvious that the co-precipitation via oxalate route, as well as the modified Pechini procedure are more favorable to drive towards high-quality, fine, single phase, and microstructurally homogeneous $Ba(Ti,Zr)O_3$ nanopowders, than the sol-precipitation method described here.

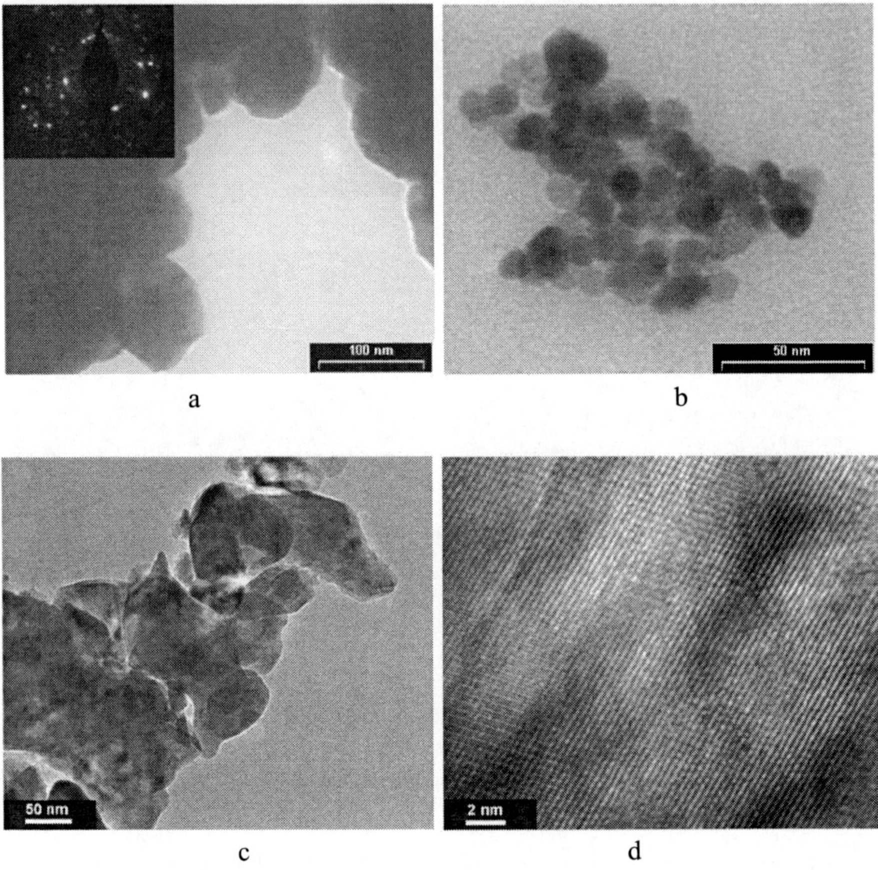

Figure 12. Images of $Ba(Ti,Zr)O_3$ powders synthesized by: (a) sol-precipitation route (TEM - bar = 100 nm; inset - SAED pattern); (b) oxalate route (TEM - bar = 50 nm); (c) modified Pechini method (TEM - bar = 100 nm) and (d) modified Pechini method (HRTEM - bar = 2 nm).

3. PREPARATION AND CHARACTERISTICS OF BA(TI,ZR)O₃ CERAMICS: EFFECT OF COMPOSITION AND GRAIN SIZE ON THE FUNCTIONAL PROPERTIES

As already mentioned, composition and grain size driven ferroelectric – relaxor phase transitions are expected in $BaTiO_3$ – $BaZrO_3$ system. While data regarding the influence of the composition *i.e.* zirconium content on the ferroelectric – relaxor crossover were reported in several scientific papers [34, 37, 66, 72], the effect of the grain size on the ferroelectric behavior of the $Ba(Ti,Zr)O_3$ solid solutions was less studied [11, 65]. However, it was suggested that for a fixed zirconium concentration, the smaller the ceramic grains, the more favorable the relaxor state. Therefore, in order to obtain such microstructurally controlled, fine-grained (submicron) BTZ ceramics, three approaches are taken into account: (*i*) establishing of suitable classical sintering strategies by a carefully optimization of sintering parameters in the case of the solid state reaction processing; (*ii*) use of nanopowders prepared by wet-chemical methods; (*iii*) use of alternative non-conventional sintering procedures able to induce grain growth inhibition by avoiding diffusional processes.

3.1. CERAMICS FROM POWDERS PREPARED BY CLASSICAL SOLID STATE REACTION

3.1.1. Phase Composition and Microstructure

The first attempt in finding effects related to grain size and compositions around the pinched phase transition was to use powders produced by classical solid state and to control the microstructures by the sintering parameters. A single high temperature-processing step to obtain $BaTi_{1-x}Zr_xO_3$ mixed crystals with compositions $x = 0$, 0.10, 0.15 and 0.18 was employed [62, 63]. To demonstrate features related to the ferroelectric-relaxor crossover induced by reduction the grain size, a ferroelectric composition was chosen ($x = 0.10$). For this composition, ceramics with a few grain sizes down to submicron values were obtained.

High-purity starting nanopowders of: $BaCO_3$ (Solvay), TiO_2 (Tosoh) and ZrO_2 (Tosoh) were wet-mixed with distilled water for 48 h. The degree of mixing for various additions of polyacrylic acid (PAA) and after different mixing times was checked by Scanning Electron Microscopy (SEM). After freeze-drying, the powders were calcined at 1000°C for 6 hours to promote the solid-state reaction. The phase formation after calcination was analyzed by X-ray diffraction. The calcined powders were sieved and manually re-milled, then compacted in cylinders (length $\sim 2 - 3$ cm, diameters of ~ 1 cm) by cold isostatic pressing at 1500 bar and then sintered at different temperatures (1350 $- 1500$°C) for 2 hours. After sintering, the density was measured, the phase purity was determined by X-ray diffraction and the microstructures were examined by SEM. The ceramic samples were then cut, polished and deposited electrodes on the plane-parallel surfaces for the electric characterization (dielectric and ferroelectric experiments).

Figure 13 shows the XRD pattern of the calcined powders, for different compositions of Zr: $x = 0.10$, 0.15 and 0.18. The XRD analysis proved that after calcination at 1000°C for 6 h all the compositions show the presence of the major titanate phase with cubic or pseudo-cubic symmetry and some wt % of orthorhombic $BaCO_3$ (see the features at 2θ-$CoK\alpha \sim 51°$).

Slow step scanning XRD analysis performed for a few calcination temperatures revealed that the formation of $BaTiO_3$ phase starts at ~ 700°C while of $BaZrO_3$ at ~ 800°C. The XRD patterns also suggest that the rate of $BaZrO_3$ formation is lower than that of $BaTiO_3$. After two hours of sintering the $BaZr_xTi_{1-x}O_3$ pellets at a few temperatures (1350°C, 1400°C, 1450°C and

1500°C), the pure perovskite phase was fully formed (figure 14). Thus, above 1350°C, BTZ phase formation increases rapidly due to the inter-diffusion between $BaTiO_3$ and $BaZrO_3$ components.

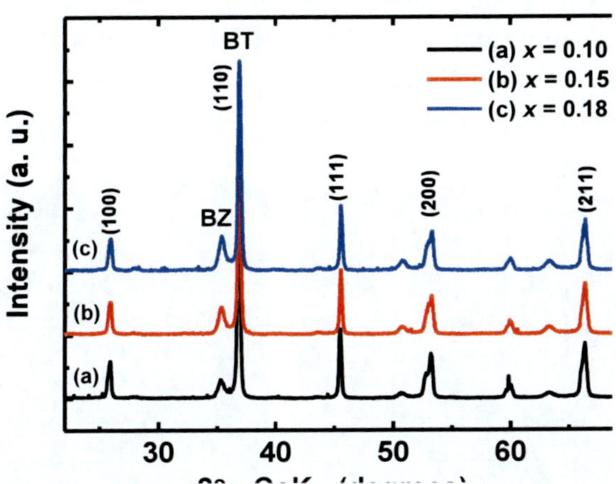

Figure 13. Room temperature XRD patterns of the powders calcined for 6 h at 1000°C for different compositions.

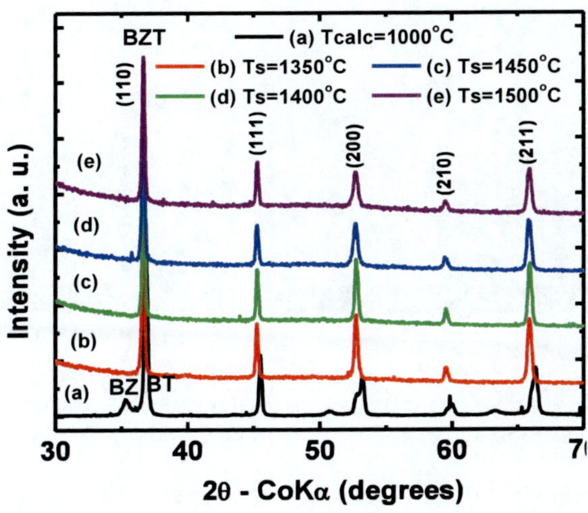

Figure 14. XRD patterns of calcined powder and of the $BaTi_{0.9}Zr_{0.1}O_3$ ceramics sintered at 1350°C, 1400°C, 1450°C and 1500°C. The secondary phases, if present, are below the XRD detection limit in all the sintered samples.

Figure 14 shows that the sample sintered at 1350°C has a cubic symmetry. With increasing the sintering temperature, the ceramic goes toward a tetragonal symmetry. The (200) peak became wider (2θ-CoKα ~ 53°) and decreased in height, which indicates the tetragonal state and thus, ferroelectric character for the ceramic with composition $x = 0.10$. A few SEM images of the BaTi$_{1-x}$Zr$_x$O$_3$ ceramics with $x = 0.10$ after sintering are shown in figure 15. The SEM analysis shows a good homogeneity of the microstructures and a certain degree of open (inter-grain) and closed (intra-grain) porosity.

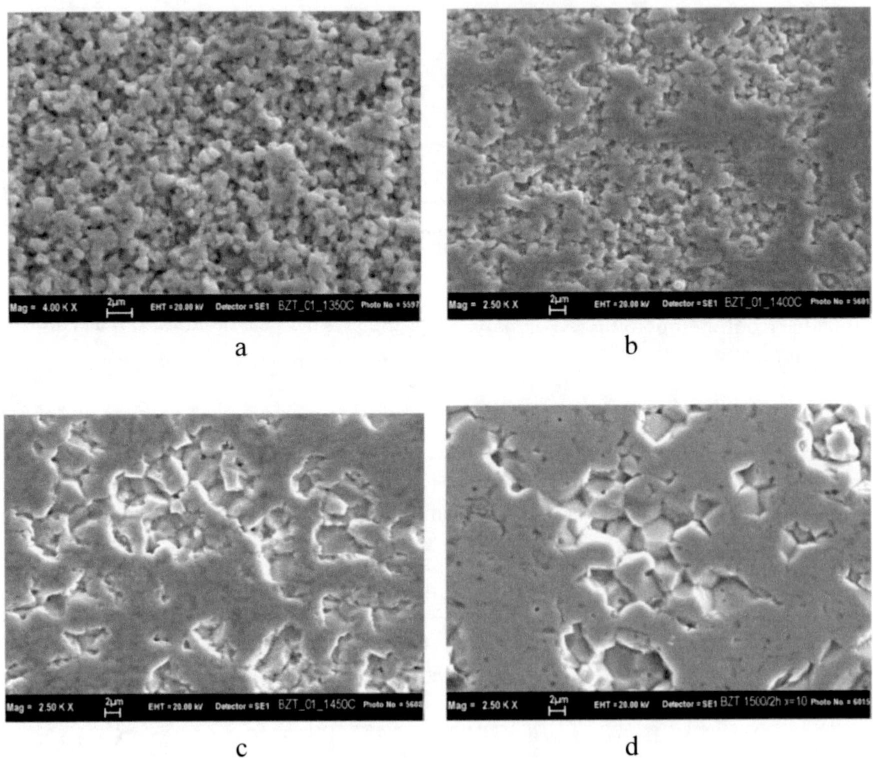

Figure 15. SEM micrographs of BaTi$_{0.9}$Zr$_{0.1}$O$_3$ ceramics (fractured surfaces) after sintering at: (a) 1350°C, (b) 1400°C, (c) 1450°C and (d) 1500°C.

The grain size of the BTZ samples after sintering, as visible after chemical attack, was determined from the SEM photos presented in the figure 16, by linear intercept method (Table 2). As expected, the average grain size of BTZ ceramics with $x = 0.10$ increases with the sintering temperature (1350°C, 1400°C, 1450°C and 1500°C).

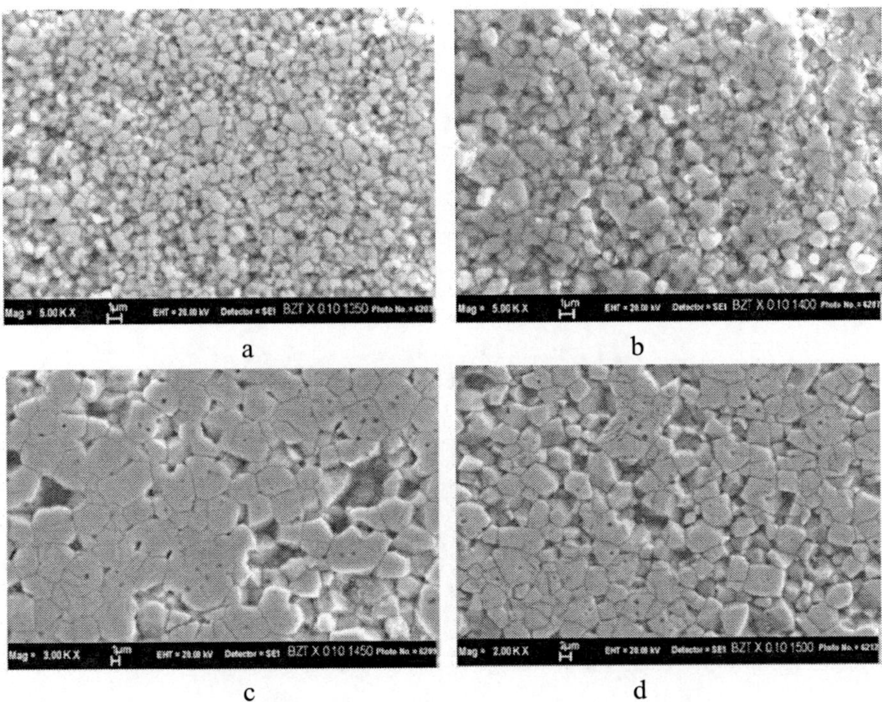

Figure 16. SEM micrographs of BaTi$_{0.9}$ Zr$_{0.1}$O$_3$ ceramics after sintering at: (a) 1350°C (bar = 1 μm); (b) 1400°C (bar = 1 μm); (c) 1450°C (bar = 1 μm) and (d) 1500°C (bar = 2 μm).

At a given sintering temperature (1500°C), the grain size is as smaller as Zr content is higher (3.27 μm, 1.8 μm and 1.54 μm for x = 0.10, 0.15 and 0.18, respectively) (figure 17). The decrease in grain size with the increase in Zr content is associated with the lower grain-growth rate due to the slow diffusion of the Zr^{4+} ion, which has a bigger radius than Ti^{4+}.

Table 2. Characteristic of BaTi$_{1-x}$Zr$_x$O$_3$ ceramic samples after sintering

Sample	Temperature of calcination (°C)	Temperature of sintering (°C)	Sintered Density (%)	Grain size (μm)
BaTi$_{0.9}$ Zr$_{0.1}$O$_3$	1000	1350	89.40	0.75
		1400	90.70	1.10
		1450	98.40	2.62
		1500	99.00	3.27
BaTi$_{0.85}$Zr$_{0.15}$O$_3$	1000	1500	89.60	1.80
BaTi$_{0.82}$Zr$_{0.18}$O$_3$	1000	1500	92.13	1.54

 Adelina Ianculescu and Liliana Mitoseriu

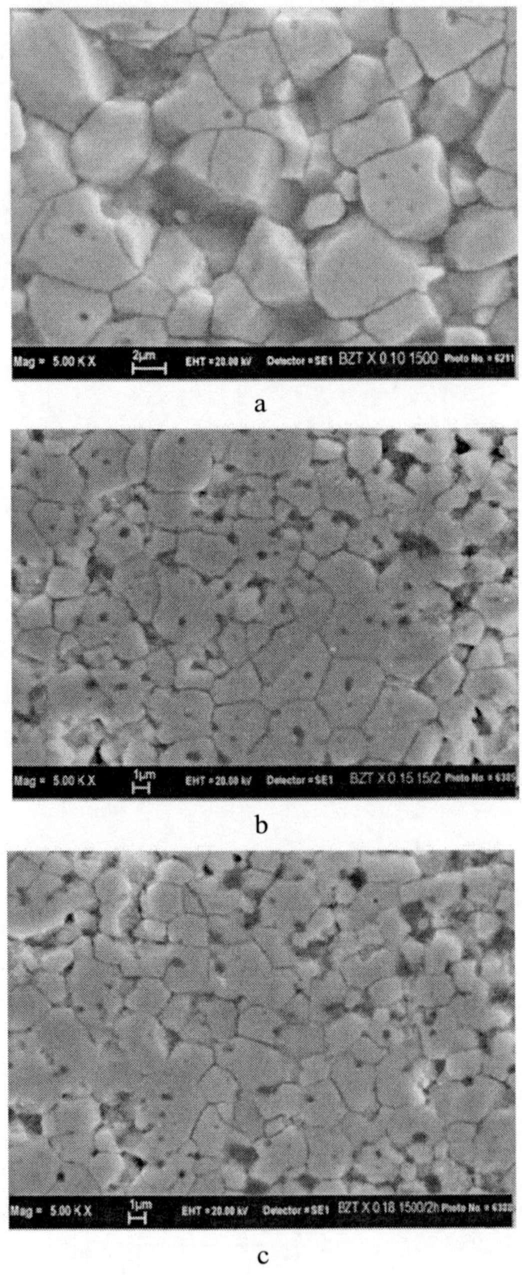

Figure 17. SEM micrographs of BaTi$_{1-x}$Zr$_x$O$_3$ ceramics with different content of Zr: (a) $x = 0.10$ (bar = 2 μm); (b) $x = 0.15$ (bar = 1 μm), and (c) $x = 0.18$ (bar = 1 μm).

3.1.2. Dielectric Properties

For the study of the dielectric properties as a function of temperature and frequency, it was used an Impedance Spectroscopy system (Solartron, SI 1260) [98]. The Impendence Spectroscopy method allows to determine not only the rough dielectric properties, but if present, to find characteristics derived by the presence of different dielectric and conductive properties in some regions of the sample, like for example grain boundary-bulk, electrode-ceramic interfaces, regions with accidental compositional inhomogeneity if present, etc. [99-101]. The method was applied in order to determine the dielectric constant and losses of our BaTi$_{1-x}$Zr$_x$O$_3$ ceramics and to check the presence of the relaxor behavior. The results are presented in the following.

The classical method to check the existence of the ferroelectric-paraelectric phase transition is to plot the temperature dependence of the permittivity. As a demonstration for the relaxor properties, typical frequency dispersion (relaxation) with a frequency shift of the temperature corresponding to the maximum in the ferroelectric state should be observed. On the other hand, the tangent loss gives information on the dielectric quality of the prepared samples. From the point of view of applications for MLCC or as tunable element, a good dielectric should have tan δ < 5% at the temperature needed for that application. A sample with losses above unity, which is normally accompanied by high apparent permittivity, can not be considered anymore a "dielectric". However, even such samples might find interesting applications. In any case the real and imaginary parts of the permittivity (or the tangent loss) have to be discussed together.

(a) The Effect of Frequency on the Dielectric Data for the BaTi$_{0.9}$Zr$_{0.1}$O$_3$ Ceramics Sintered at Different Temperatures (Size Effects)

Figures 18, 19 and 20 present the real part of the permittivity and losses *vs.* temperature, at a few selected frequencies for the BaTi$_{0.9}$Zr$_{0.1}$O$_3$ ceramics, sintered at various temperatures (1350°C – 1500°C), thus having various grain sizes in the range of 0.75 – 3.27 μm (Table 2). The permittivity *vs.* temperature dependence presents a maximum located at low frequency at the temperature T_m ≈ 90°C, which shifts towards higher values when the frequency increases. In the ferroelectric phase (for $T < T_m$), the permittivity of all the ceramics presents a relaxation in the range of 1 – 0.5 × 10^5 Hz, with a shift of T_m of 3 – 5°C (figures 18 (a), 19 (a) and 20 (a)). This behavior demonstrates that the present ceramics have typical relaxor character. It can be also observed a clear tendency to reduce the relaxation as the average grain size increases: the coarse ceramic sintered at 1500°C shows almost no relaxation and its dielectric properties are close to the typical ferroelectric ones (figure 20 (a)). In fact, in the literature, the composition $x = 0.10$ was reported to be fully ferroelectric [10, 12, 13, 34-37]. This fact might be true for single-crystal and coarse ceramics. In the case of the present ceramic samples, the coarsest one is close to such a character, because it presents a sharper ferroelectric-paraelectric phase transition and only a very small frequency relaxation in the polar phase.

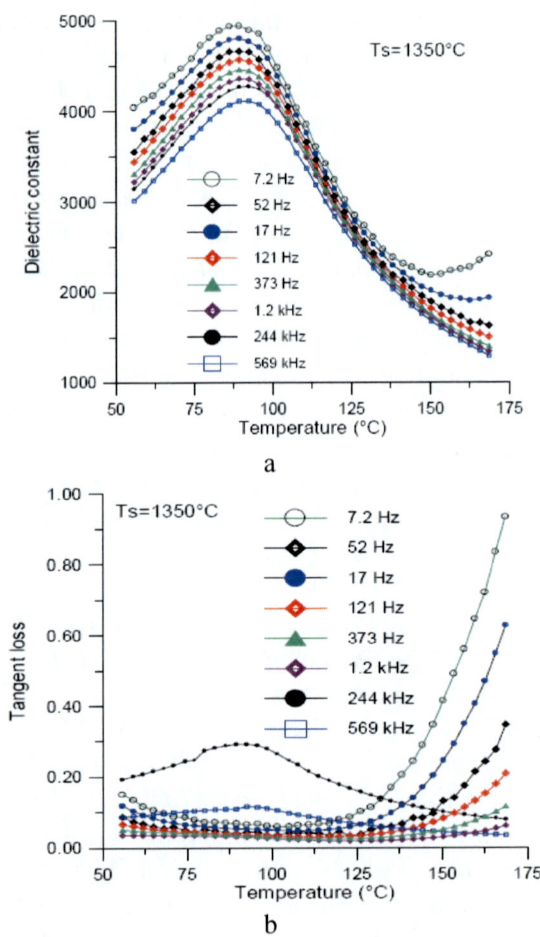

Figure 18. (a) Permittivity and (b) tangent loss *vs.* temperature for the $BaTi_{0.9}Zr_{0.1}O_3$ ceramics with average grain size of 0.75 μm.

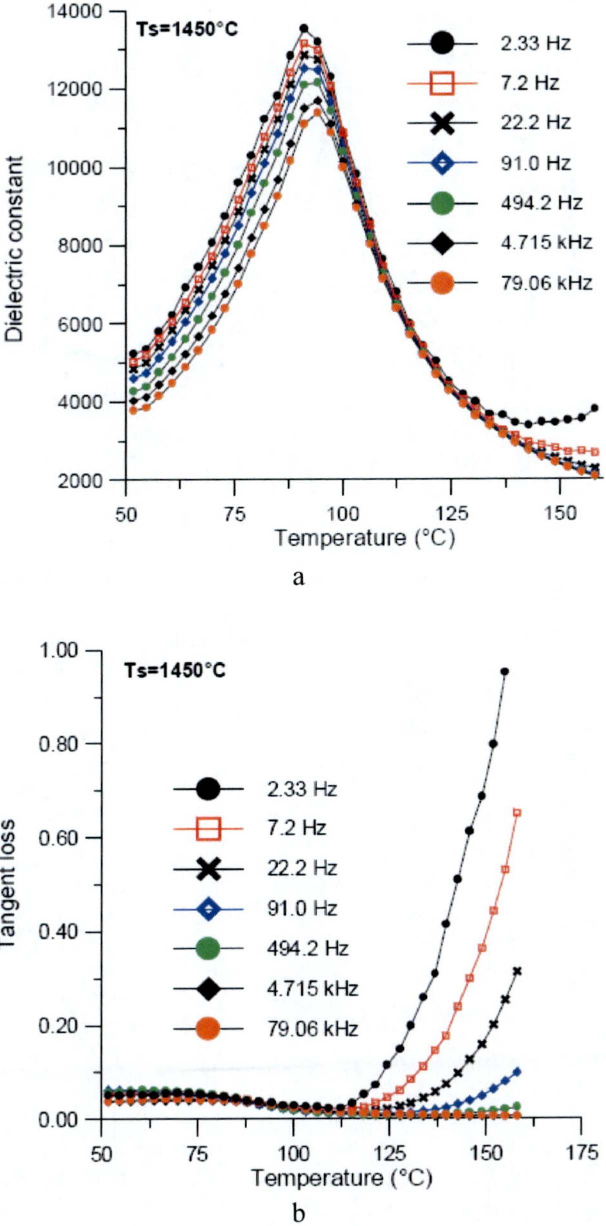

Figure 19. (a) Permittivity and (b) tangent loss *vs.* temperature for the BaTi$_{0.9}$Zr$_{0.1}$O$_3$ ceramics with average grain size of 2.62 μm.

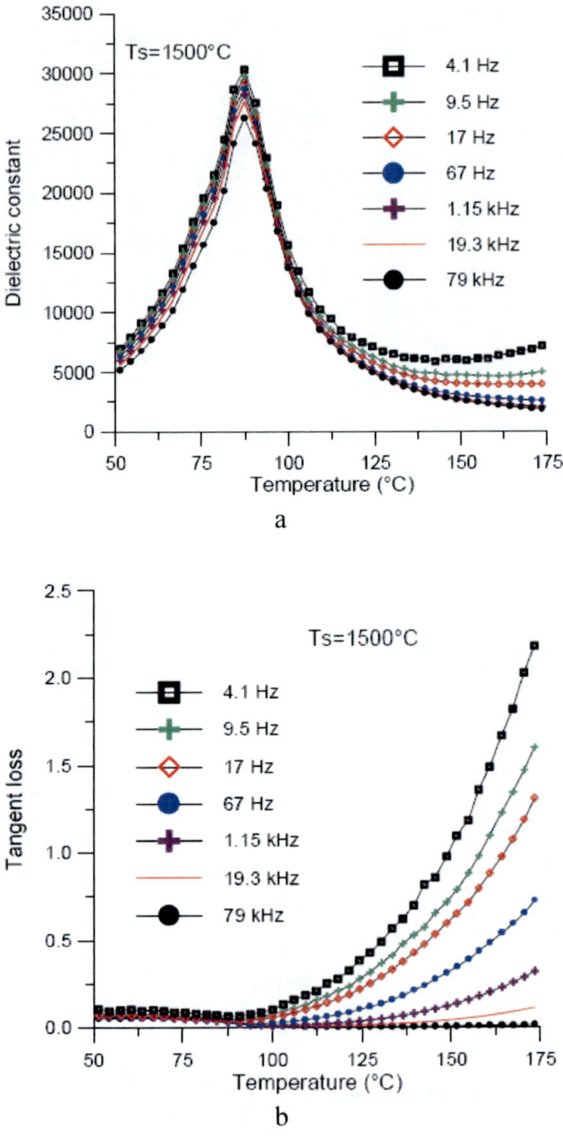

Figure 20. (a) Permittivity and (b) tangent loss *vs.* temperature for the $BaTi_{0.9}Zr_{0.1}O_3$ ceramics with average grain size of 3.27 μm.

Usually, the diffuse character of the phase transition in relaxors was described by various types of empirical laws which can be rigorously deduced within models describing the relaxor state [102-106]. One of the most used proposed in the Ref. [107] has the advantage to describe the permittivity-temperature dependence in relaxors even in the dispersion region, by using unique parameters for all the frequencies:

$$\varepsilon = \frac{\varepsilon_m}{1+\left(\dfrac{T-T_m}{\delta}\right)^{\eta}} \tag{5}$$

where ε is the dielectric constant at a given temperature, ε_m is its maximum value at the corresponding temperature T_m, δ indicates the range of temperature extension for the diffuse phase transition owing to the direct correlation with the dielectric permittivity broadening. The parameter η gives information on the character of the phase transition: for $\eta = 1$, normal Curie-Weiss law is obtained while $\eta = 2$ describe a complete diffuse phase transition (full relaxor character). For a ferroelectric material, eq. (5) reduces to the Curie-Weiss law, with $\eta = 1$ and in this situation δ is proportional to the Curie constant.

By fitting the dielectric data of the samples with extreme grain sizes sintered at 1350°C and 1500°C with the empirical eq. (5) at the frequency of 79 kHz, it was confirmed that the relaxor character is increasing as the ceramic grain size decreases:

- the finest ceramic (grain size of 0.75 μm) has $\eta = 1.7$ and $\delta = 46$°C by comparison with
- the coarse ceramic (grain size of 3.27 μm), with $\eta = 1.5$ and $\delta = 20$°C.

The results of these fits show that even the coarsest ceramic is not fully ferroelectric, but it has some degree of relaxor contribution.

By this analysis it seems that the relaxor state is favored when reducing grain size (grain size-induced ferroelectric-relaxor crossover). The present result is in agreement with the recent ones reported in ref. [43], in which was shown that BTZ ceramics with the composition $x = 0.20$ prepared by sol-gel exhibited a similar grain size-induced crossover. In the present case, the composition $x = 0.10$ showing similar trend is even closer to the full ferroelectric state. It seems that appropriately reducing the grain size of the ceramics, even without chemical heterogeneity (*i.e.* in pure ferroelectrics like for example BaTiO$_3$) is possible to reach the relaxor (superparaelectric) state, if no other stronger effects (like grain boundary phenomena) are overlapping to this intrinsic tendency of the system. This possibility was recently investigated for pure $BaTiO_3$ ceramics, where the reduction of the grain size at nanoscale (down to 30 nm) also showed such a trend [108-111]. It seems that if the ferroelectric system is "pushed" towards disorder by chemical substitution (here Zr addition in the B-site of the perovskite cell), not so small grain size (below $1\,\mu m$) are enough to induce a ferroelectric-relaxor crossover. Both the effect of chemical heterogeneity and of reducing grain size are contributing in changing the balance between long range order/short range order in the BTZ system, and this acts in shifting the system towards one of the two behaviors.

Usually in ceramics, particularly in solid solutions, the local composition inhomogeneity gives rise to local variation of the polarization resulting in uncompensated bounded charges $\rho_{pol} = -\mathrm{div}\vec{P}$ in the volume of ceramic sample. These normally produce an extrinsic electrostatic polarization (not related to the ferroelectric lattice), called *space charge* or *Maxwell-Wagner* effect [35, 41]. The present BTZ solid solutions also show such a phenomenon, which manifests by a frequency-dependence of the dielectric constant together with an anomalous increase of the dielectric losses at high temperatures, as observed in figures for all samples in figures 18, 19 and 20. This phenomenon is not related with relaxor or ferroelectric states and it was found in many inhomogeneous perovskite-like systems. Its relaxation time is rather high, so that it manifests particularly at low frequency (generally below 100 Hz). At higher frequency this contribution disappears and is completely missing at about 500 kHz.

The dielectric losses (figures 18 (b), 19 (b) and 20 (b)) demonstrate good dielectric properties of the $BaTi_{0.9}Zr_{0.1}O_3$ ceramics. At room temperatures and below 110°C, the losses are smaller than 20% for all measured frequencies and below 5% in the kHz range. A strong increase of the losses above unity for temperatures above 125°C, particularly at low frequencies (Hz) correlated with the apparent increasing of the dielectric constant, confirms that a thermally-activated relaxation of *Maxwell-Wagner* type affects the dielectric properties in this range of temperatures. This type of phenomenon is more evident in [35, 41]: *(i)* solid solutions than in pure materials, *(ii)* in grained structures (ceramics and films) than in single-crystals. In the case *(i)* the local chemical heterogeneity causes local variations of the polarization and thus, produces uncompensated volume charges, while in the situation *(ii)*, the interfaces are regions where the polarization is strongly inhomogeneous, again resulting in uncompensated surface charges. These *space charges* are bonded (mobile ionic species or electrons, according to the mechanism involved to maintain the electrical neutrality) and possess a certain mobility under input electric field, but only in confined outer or inner interfaces. The space charge is contributing to the total polarization of the ferroelectric ceramic and to its dielectric response, as it was observed for the present BTZ ceramics.

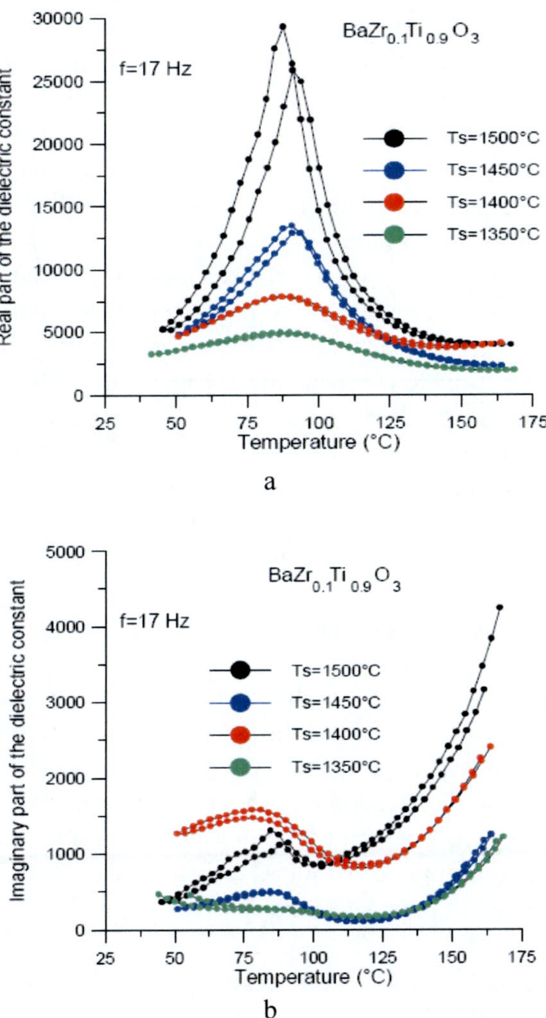

Figure 21. Grain size dependence of the: (a) real part and (b) imaginary part of the dielectric constant *vs.* temperature for the BaTi$_{0.9}$Zr$_{0.1}$O$_3$ ceramics with various grain sizes.

The grain size effect on the dielectric characteristics can be observed in the figure 21, at a fixed frequency. The permittivity of all the ceramic samples with $x = 0.10$ assumes very high values at room temperatures, of around 3000 – 4000 for the finest (and less dense) ceramic sintered at 1350°C and of 5000 – 7500 for the coarse ones sintered at 1500°C. At the transition point T_m, the permittivity reaches very high values of ~30,000 for the coarse sample, which is a result of the combined high and homogeneous grain size and the highest densification degree obtained in these ceramics. These properties demonstrate that by improving the microstructure, excellent dielectric properties can be found for a given composition.

Figure 21 shows how the real and imaginary components of the dielectric constant are dependent on grain size in a heating/cooling cycle, which is accompanied by thermal hysteresis that increases with increasing grain size. On the coarsest ceramic (which is excellent from the dielectric and ferroelectric point of view due to its high density and large homogenous grains), the temperature dependence of the imaginary part allows to identify both the tetragonal – cubic (ferroelectric – paraelectric) phase transition at ~ 90°C, and also the structural orthorhombic – tetragonal (ferroelectric – ferroelectric) transition at ~ 75°C by a small anomaly (figure 21 (b)). These transitions are convoluted for the finest ceramic. The fact that the dielectric constant is so high in the temperature range of ~ 70°C – 100°C is a result of proximity of both the structural phase transitions.

(b) The Role of Composition on the Dielectric Properties of BaZr$_x$Ti$_{1-x}$O$_3$ Ceramics

The role of composition on the dielectric properties of BaTi$_{1-x}$Zr$_x$O$_3$ ceramics which were sintered at the same temperature can be seen by comparing the figures 20, 22 and 23. As expected, the relaxor character is favored by increasing the Zr addition. This is also demonstrated by the results of fits with the empirical eq. (5) of the dielectric data obtained for BTZ with various compositions, sintered at 1500°C (at a fixed frequency of 79 kHz). The values of the resulted fit parameters are:

- $\eta = 1.5$ and $\delta = 20°C$ for $x = 0.10$;
- $\eta = 1.65$ and $\delta = 32°C$ for $x = 0.15$;
- $\eta = 1.7$ and $\delta = 36°C$ for $x = 0.18$.

These values indicate a mixed ferroelectric-relaxor for all the compositions and an increasing relaxor character with Zr addition. The parameter δ indicates the temperature extension for the diffuse phase transition, that is correlated with the dielectric permittivity broadening and shows the gradual increasing of the diffuse character of the phase transition with increasing the Zr concentration.

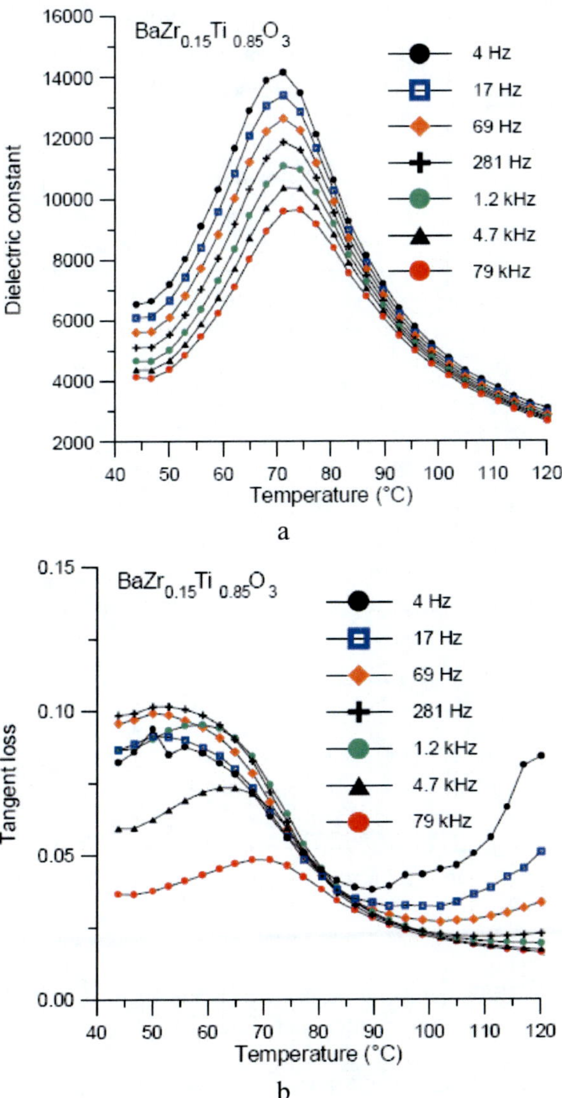

Figure 22. Real part of the dielectric constant (a) and tangent loss (b) *vs.* temperature for the BaTi$_{0.85}$Zr$_{0.15}$O$_3$ ceramics, sintered at 1500°C at a few frequencies.

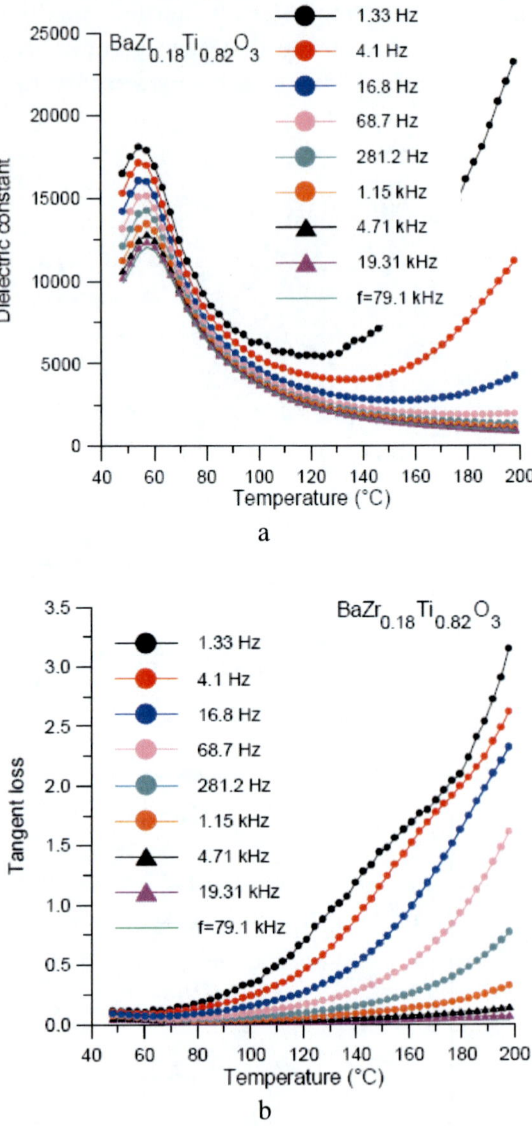

Figure 23. Real part of the dielectric constant (a) and tangent loss (b) *vs.* temperature for the BaTi$_{0.82}$Zr$_{0.18}$O$_3$ ceramics, sintered at 1500°C at a few frequencies.

Remarkable low losses together with a very high dielectric constant of 16,000 at $T_\mathrm{m} \approx 70°C$ are observed for the sample with composition $x = 0.15$ (tan $\delta <$ 5% at 80 kHz and tan $\delta <$ 10% for all frequencies in the temperature range of 40 − 120°C). This ceramic also presents a negligible space charge effect (figure 22 (b)). This sample is the only one to strictly follow a modified Curie-Weiss dependence in the paraelectric state. For some reasons, a higher degree of the electrical homogeneity was obtained for this composition. The ceramic with the composition $x = 0.18$ has higher losses (in any case smaller than 10% at room temperature even at the lowest frequency $f = 1.33$ Hz) and a strong space charge contribution, giving rise to losses above unity for $T >$ 130°C (figure 23 (b)). The maximum is located at $T_m \approx 60°C$ and the dielectric constant at this temperature also reaches high values, of 18,000 (figure 23 (a)).

The composition-induced shift of the ferro-para phase transition at a fixed frequency $f = 80$ kHz is visible in the figure 24, where the temperature-dependence of the dielectric constant of the BTZ solid solutions by comparasion with a normal BT ceramic prepared in the same condition is reported. The difference between the sharp ferro-para phase transition in BT (fully ferroelectric) and the diffuse phase transitions in the solid solutions, that for any composition present a mixed ferroelectric-relaxor behavior is visible in this figure.

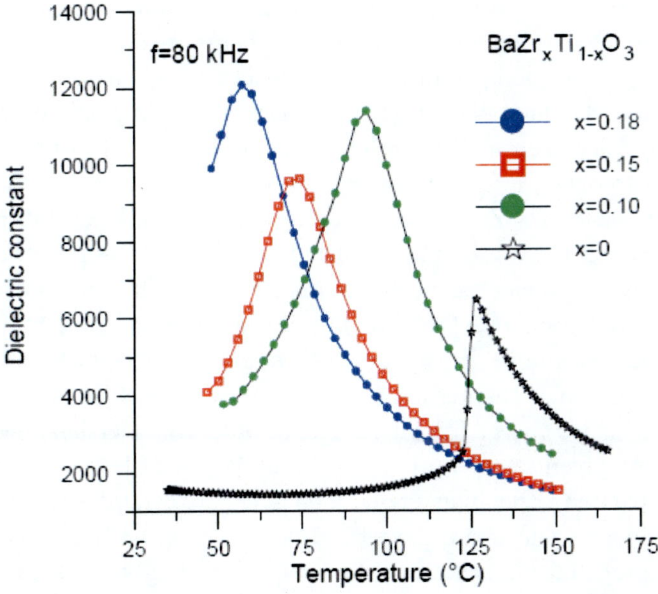

Figure 24. Real part of the dielectric constant vs. temperature for the BaTi$_{1-x}$Zr$_x$O$_3$ ceramics with $x = 0$, 0.10, 0.15 and 0.18 at a fixed frequency, illustrating the compositionally-induced shift of the Curie temperature.

(C) Frequency-Dependence of the Dielectric Constant

The frequency-dependence of the dielectric constant in the $BaTi_{0.9}Zr_{0.1}O_3$ ceramics with different grain sizes at a few temperatures including the transition range is comparatively presented in figure 25 (a-d). All samples present high values of permittivity at high temperatures and low frequencies, below 1 Hz, followed by a monotoneous decreasing as the frequency is increasing and conserving the temperature-dependence due to the ferro-para phase transition (*i.e.* an increase up to $T_m \approx 90°C$ followed by a modified Curie-Weiss reduction at higher temperatures). In this frequency region the space charge phenomena are causing a temperature-induced conductivity. The lack of anomalies in the dielectric spectra for the samples sintered at 1350°C, 1450°C and 1500°C (figure 25 (a), (c) and (d)) are proving that no other relaxation phenomena are active except the mentioned one, besides the lattice contribution that has a normal reduction of permittivity with frequency. The samples with compositions $x = 0.15$ and $x = 0.18$ (not shown here) present a normal relaxation like the ceramics $x = 0.10$ sintered at 1350°C, 1450°C and 1500°C. The completely different behavior of the sample sintered at 1400°C (figure 25 (b)) has the origin in the presence of another thermally-activated relaxation process in addition to the space charge mechanism giving rise to conductivity at high temperatures. This process does not causes conductive loss and is active in the range 100 Hz − 10 kHz, depending on the temperature, as proved by the presence of the inflexion points in this range of frequency.

In order to obtain more information on the possible origin of the observed behavior of the $BaTi_{0.9}Zr_{0.1}O_3$ ceramics sintered at 1400°C, the Impedance Spectroscopy method was employed. The impedance analysis has been widely used to study the dielectric behavior of the crystalline and polycrystalline ceramic materials [99, 112-115]. In general, the dielectric properties of ferro-electromagnetic ceramics arise due to intra-grain, inter-grain and electrode effects. The motion of charges could take place by charge displacement, dipole reorientation, space charge formation, etc. In order to understand the electrical properties of a given sample grain, grain boundary and electrode contributions must be separated. The inter-grain boundaries in ceramics are a defective region (oxygen vacancies are the typical defect in perovskites) as well as the place of segregated impurities or dopants. Accordingly, it is expected that from the electrical point of view, the grain bulk and grain boundary regions have different dielectric and conductive properties. In addition, the ceramic region in contact with electrodes **might be "modified" with respect to the inner** part of the bulk: different Fermi levels of the two materials in contact give rise to Schottky barriers, imperfect conductivity of the electrodes create a screening region changing the potential distribution in their neighborhood and metallic ions might diffuse into the ceramic creating a doped interface regions. In any of these situations, the dielectric and conductive properties at the contact electrode-ceramics might be different with respect to the rest of material.

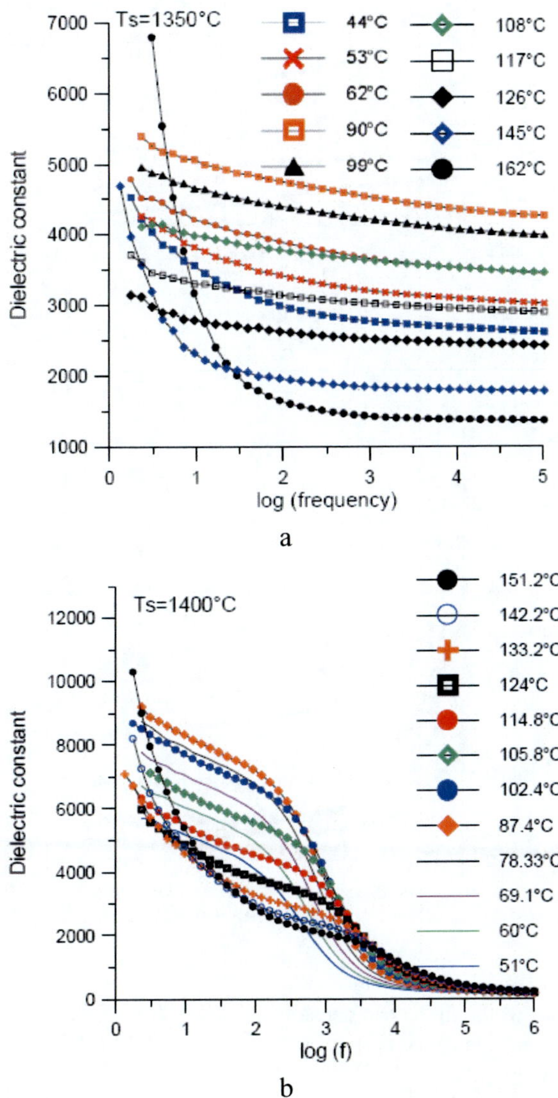

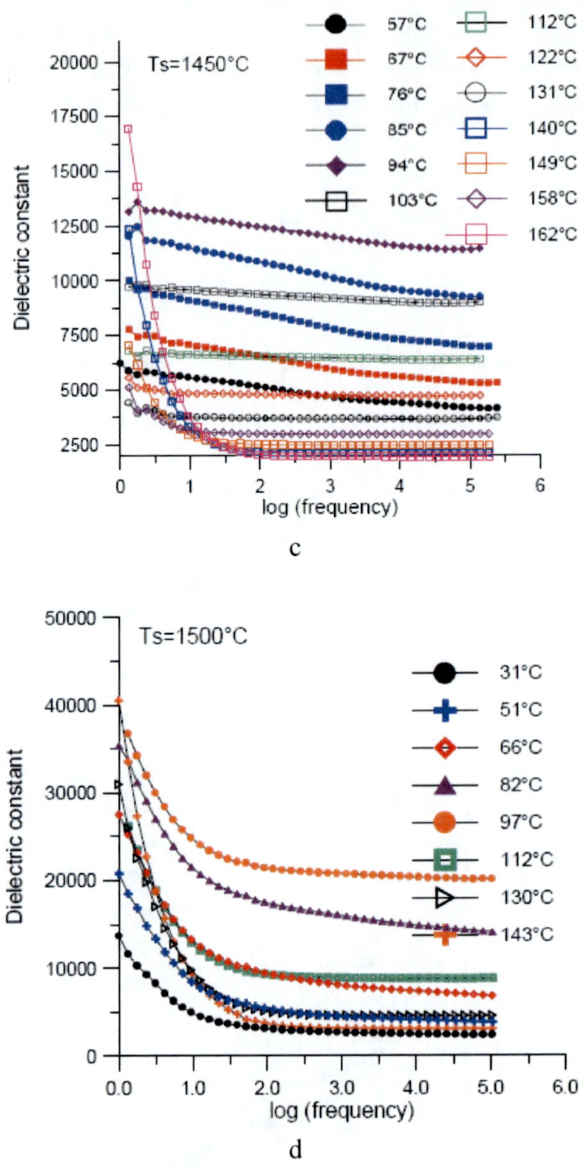

Figure 25. Real part of the dielectric constant *vs.* logarithm of frequency for the BaTi$_{0.9}$Zr$_{0.1}$O$_3$ ceramics, sintered at: (a) 1350°C; (b) 1400°C; (c) 1450°C and (d) 1500°C, at a few temperatures.

The complex impedance analysis has proved as a very powerful tool for separating these contributions and it was used here in order to check if such kind of inhomogeneities give contribution to the complex impedance spectra of the BTZ ceramics.

A homogeneous dielectric should be described by a Debye eq. [116]:

$$\varepsilon^* - \varepsilon_\infty = \frac{\varepsilon_s - \varepsilon_\infty}{1 + i\omega\tau},\tag{6}$$

Cole-Cole [117]:

$$\varepsilon^* - \varepsilon_\infty = \frac{\varepsilon_s - \varepsilon_\infty}{(1 + i\omega\tau)^{1-\alpha}},\tag{7}$$

or more generally, by the *universal relaxation power-law* as proposed by Jonsker [118]:

$$\varepsilon^* = \frac{\varepsilon_s + \varepsilon_\infty (i\omega\tau)^q}{1 + (i\omega\tau)^q},\tag{8}$$

where:

$$\varepsilon^* = \varepsilon' - i\varepsilon''\tag{9}$$

is the complex permittivity, ε_s is the static component (at zero frequency) and ε_∞ is the permittivity at very high frequency and α and q non-dimensional parameters. In terms of the complex dielectric constant representation (ε'' *vs.* ε') or impedance representation (Z'' *vs.* Z'), an ideal simple Debye relaxation will be represented as a semicircle in the complex plane. The data for real homogeneous dielectrics actually fit an arc of a circle with the center below the real axis [118] and is described by an RC-equivalent circuit.

For a polycrystalline ceramic material with electrodes, a more complex model is appropriate, in which different groups of RC-circuits describing: (a) the ceramic grain, (b) the ceramic grain boundaries, (c) the electrode-ceramic interfaces (figure 26). The last one gives a very important contribution in thin film structures, because the "modified" ceramic at the interface with electrode occupies an important volume ratio of the sample. The second component is missing in single-crystals and is enhanced in nanosized ceramics, due to the large number of grain boundaries in such structures. Depending on their electrical properties (values of the equivalent impedance of the regions), the contributions are often convoluted and in such a case their separation is very difficult. When the electrode contribution is negligible, the complex impedance plot contains two arches of circles corresponding to the grain boundary and bulk.

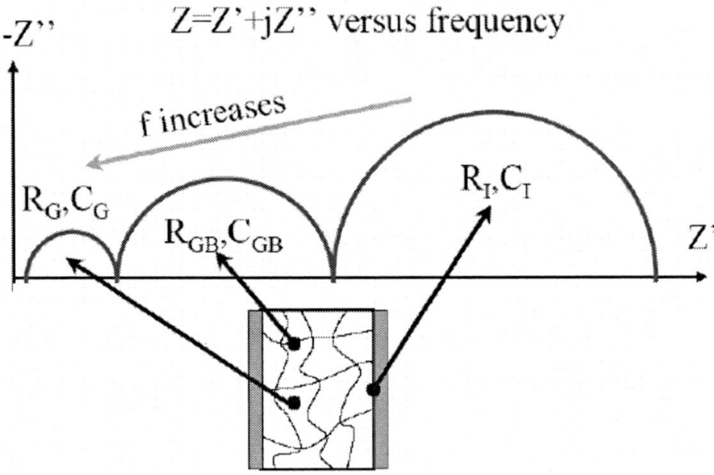

Figure 26. Model adopted for a real electroded ceramic containing three distinct contributions: ceramic grain (R_G, C_G), ceramic grain boundaries (R_{GB}, C_{GB}) and electrode-ceramic interfaces (R_I, C_I).

Figure 27 shows the complex impedance plots obtained at two temperatures for the BaTi$_{0.9}$Zr$_{0.1}$O$_3$ ceramics, sintered at 1350°C, 1400°C and 1500°C. All these impedance diagrams start from the origin, showing that the material does not have a dc-conductivity (which can be represented as a series resistor to the parallel group of figure 26), thus the present materials are really good dielectrics. The complex impedance plots are represented by a fragment of arch and they have a single component. Only the sample sintered at 1400°C clearly shows two separated components (figure 27 (b)): a small arch characterized by a bulk resistivity of around 1 MΩ and a larger radius arch resulting in a much higher grain boundary resistance. Above the ferroelectric-paraelectric phase transition, almost a single component is obtained, due to the strongly decreasing of permittivity in the paraelectric state.

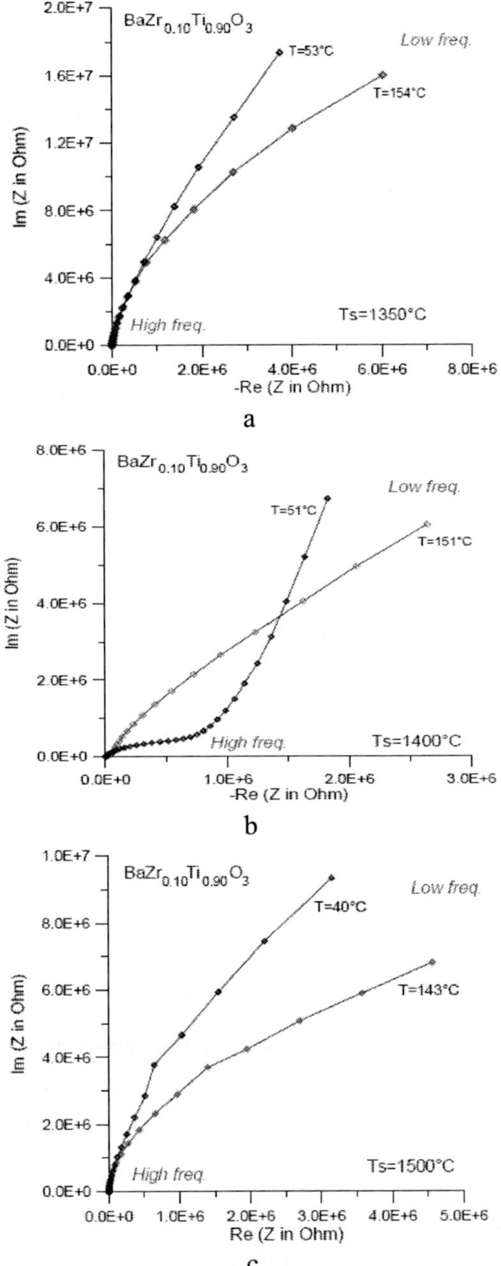

Figure 27. Complex impedance plots for the BaTi$_{0.9}$Zr$_{0.1}$O$_3$ ceramics, sintered at the temperatures: (a) 1350°C; (b) 1400°C and (c) 1500°C.

By considering the possible defects in the perovskite ceramics, it seems that the oxygen vacancies are the most probably responsible for the observed effect [35]. Their concentration might be different in the regions of grain bulk and grain boundaries as result of the high temperature sintering. It is well known that in the perovskite structure materials containing Ti, the ionization of the O-vacancy creates conducting electrons:

$$V_O \Leftrightarrow V_O^{'} + e'$$
$$V_O^{'} \Leftrightarrow V_O^{''} + e' \tag{10}$$

or these electrons might be bonded to Ti^{4+} in the form of:

$$Ti^{4+} + e' \Leftrightarrow Ti^{3+}. \tag{11}$$

As indicated by Ihrig & Hennings [119], it is difficult to determine whether the weakly bonded electrons are located near an O-vacancy or near the Ti ions. The exact location of the electrons depends on the details of structure, temperature range, etc. It was, however, shown that the O-vacancies lead to shallow level electrons. These electrons may be trapped by Ti^{4+} ions or O-vacancies and they easily become conducting electrons by thermally activation. Dielectric relaxations due to the O-vacancies were reported both in ferroelectric and non-ferroelectric perovskites like $CaTiO_3$ by Maglione *et al.* [120, 121] and more recently, by Kang *et al.* [122], in $Pb_{0.9}La_{0.1}TiO_3$, PLZT and $BaTiO_3$ ceramics. Therefore, in these materials the phenomena are activated at much higher temperatures: 600 - 800°C in pure materials and 400 - 700°C in solid solutions. The activation energy of these phenomena is reduced in defective or more inhomogeneous systems.

In the case of the present $BaTi_{0.9}Zr_{0.1}O_3$ ceramic sintered at 1400°C, the observed phenomena are present at much lower temperatures of 100 - 200°C, thus overlapping the ferro-para phase transition and the relaxor dispersion. The reason might be related to the grain boundary phenomena or to other local inhomogeneity or defects accidentally formed inside this sample during sintering. Besides the O-vacancies, local doping or small compositionally gradients resulted from temperature gradients during cooling down from sintering to room temperature might produce defects able to activate the relaxation phenomena at low temperaturs. In any case, the observed relaxation is an extrinsic property (caused by defects and grain boundary phenomena) and not by intrinsic ferroelectric/relaxor character related to the crystalline lattice, which was very similar for all the sintered solid solutions.

More recently, a similar series of BTZ ceramics were subjected to post-annealing thermal treatments at 1000°C for 50 h, allowing a reduction and re-equilibration of the O-vacancy concentration inside the ceramic body. The preliminary dielectric data showed that the giant relaxation of the BaTi$_{0.9}$Zr$_{0.1}$O$_3$ ceramic sintered at 1400°C disappeared after annealing, while the space charge (Maxwell-Wagner) phenomena were still present [123]. It results that the two observed relaxation phenomena have indeed a different nature and the giant relaxation is related to the O-vacancies. Further detailed studies are necessary to clarify all the aspects related to the dielectric relaxation mechanisms in the present ceramics. It is worth to conclude that the dielectric properties of BaTiO$_3$-based solid solutions might be affected by residual charges of various natures and by grain boundary phenomena. When the relaxor properties are claimed in a ferroelectric system on the basis of its diffuse character of the phase transition and relaxation in kHz range, one has to be sure on the type of the observed dielectric relaxation and to understand which component might be affected by extrinsic effects such as giant Maxwell-Wagner relaxations or grain boundary phenomena.

In conclusion, the performed dielectric study of the BTZ ceramics prepared by solid-state powders showed that:

- all the ceramic samples have good dielectric properties, high permittivity and rather low losses, which are dependent on the microstructure and composition;
- BTZ ceramics might present dielectric anomalies caused by extrinsic effects such as grain boundary phenomena which are produced by the insufficient re-oxidation during cooling from the sintering temperature. This phenomenon is completely removed after post-annealing thermal treatment allowing the equilibration of oxygen vacancy concentration within the ceramic body;
- BZT present more than a single defect charge mechanisms of dielectric relaxation: one was identified as due to Maxwell-Wagner phenomena (space charge effect) and one caused by the O-deficiency with different concentrations in the grain boundary and bulk;
- a diffuse character of the phase transition and ferroelectric-relaxor crossover is induced by increasing Zr addition;
- the ferroelectric-relaxor crossover for a given composition ($x = 0.10$) is favored by reducing grain size below 1 μm.

The last observation is very interesting, mostly because the superparaelectric state in ferroelectrics by reducing grain size as in ferromagnetism was never really demonstrated. However, by the dielectric study only, it is difficult to definitely conclude about the increasing relaxor character when reducing grain size, also due to possible extrinsic phenomena affecting the impedance response. An additional demonstration would be given by observing the evolution of switching properties (reducing polarization and tendency towards zero coercivity) when reducing grain size. An investigation of the polarization-field dependence in the BTZ ceramics prepared by solid-state powders is described in the following.

3.1.3. Ferroelectric Properties (Polarization-Field Responses)

The main characteristic of a ferroelectric system represents the polarization reversal *i.e.* the ability to switch its polarisation between two bi-stable states $\pm P$ under a driving field. This process is accompanied by hysteresis, which is a typical irreversible process. The aspect of the ferroelectric hysteresis loop (saturation, rectangularity) and its main characteristics: remanent polarisation P_r, saturation polarisation P_s and the coercive field E_c are typical of the material in a given structure, but they might be affected by internal parameters as impurities, defects, local inhomogeneity, etc. By resulting from an electrical measurements under high voltage, additional effects can distort the $P(E)$ loops by non-ferroelectric components created by other mechanisms as conductivity, interfacial polarization, not-perfect ohmic contacts between the ceramic and electrodes giving rise to screening, built-in bias fields, etc. Thus, the results of these measurements have to be discussed by considering such mechanisms and they have to be correlated with the dielectric data. In the following, the $P(E)$ hysteresis data obtained at room temperature at a few frequencies for the $BaZr_xTi_{1-x}O_3$ ceramics with various compositions ($x = 0.1$, 0.15 and 0.18) and for a given composition ($x = 0.10$), the effect of different grain size is presented.

The electroded BTZ ceramic pellets, as prepared for the electrical measurements were used to record the *P(E)* loops and to determine the First-order Reversal curves (FORC) based on the Preisach formalism under sinusoidal waveform of amplitude $E_0 = 1.1$ kV/mm by using a modified Sawyer-Tower circuit [98].

First-order Reversal Curves (FORC) distribution method is a new approach inspired by the Preisach modeling [124] and frequently used in the last years to describe the switching properties of various ferroics *e.g.* magnetic recording media, magnetic properties of spin glasses, ferrofluids and geological ferromagnetic systems [125], thermal hysteresis in systems with spin transitions [126] and polarization reversal characteristics of ferroelectrics [127- 130].

The FORC method involves measurements of asymmetric minor hysteresis loops obtained by cycling the sample between saturation E_{sat} and a variable reversal field $E_r \in \left(-E_{sat}, E_{sat}\right)$, according to the sequence:

I   saturation under a positive field $E \geq E_{sat}$;

II   ramping the field down to the reversal value E_r, when the polarization follows the descending branch of the Major Hysteresis Loop (MHL);

III increase the field back to the positive saturation (during this process the polarization is a function of both the actual field E and of the reversal field E_r).

The FORC family starting on the descending MHL branch is denoted as $p^-_{FORC}(E_r,\ E)$ (as represented in figure 28).

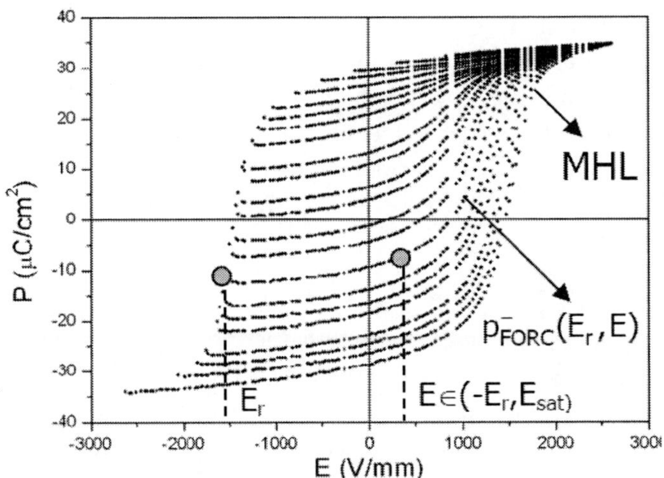

Figure 28. The definition of the First Order Reversal Curves (FORC) $\rho^-\left(E_r, E\right)$ together with the experimental MHL and a few experimental ferroelectric recorded FORCs.

In a similar way, the FORCs $p^{+}_{\text{FORC}}(E_r, E)$ can be obtained using the ascending branch of the MHL. The FORC diagram is a contour plot of the FORC distribution which is defined as the mixed second derivative of polarization with respect to E and E_r [125]:

$$\rho^{-}(E_r, E) = \frac{1}{2}\frac{\partial^2 p^{-}_{\text{FORC}}(E_r, E)}{\partial E_r \partial E} = \frac{1}{2}\frac{\partial}{\partial E_r}\left[\chi^{-}_{\text{FORC}}(E_r, E)\right], \quad (12)$$

where $\chi^{-}_{\text{FORC}}(E_r, E)$ are the differential susceptibilities measured along the FORCs. The 3D-FORC distribution $\rho^{\pm}(E_r, E)$ describes the sensitivity of polarization in a given sample with respect to changes of the reversal field E_r and actual electric field E. By changing the coordinates of the FORC distribution from (E, E_r) to $\left\{E_c = (E - E_r)/2, E_{\text{bias}} = (E + E_r)/2\right\}$, where E_c and E_{bias} play the role of local coercive and bias fields, respectively, $\rho^{\pm}(E_r, E)$ becomes a distribution of the switchable units over their coercive and interaction (bias) fields. The main advantage of the FORC method is that it is an experimental, model independent technique which is consequently applicable for describing any ferroic system. Another important advantage of the method represents the possibility to evaluate separately the reversible and irreversible contributions to the total polarization by using the same set of experimental data, together with a proper numerical procedure [131]. The reversible part of the FORC distribution, as due to the low-field contributions to the polarization is calculated as:

$$\rho^{\pm}_{\text{rev}}(E_r) = \lim_{E \to E_r, E > E_r} \rho^{\pm}_{\text{FORC}}(E_r, E) \qquad (13)$$

The reversible contribution revealed by FORCs can be compared with the results obtained by other investigations such as the sub-switching (Rayleigh) loops or the low-signal capacitance measurements [132, 133]. In the case of ferroelectrics, it was previously demonstrated that FORC diagrams are sensitive to the frequency of the applied field, crystalline orientation and fatigue effects [127, 128]. "Healthy" and homogeneous ferroelectrics with rectangular hysteresis loops are characterized by FORC distributions with a sharp maximum located at well defined fields $(E_{\text{c,M}}, E_{\text{bias,M}})$ and with almost zero reversible component. Such a sharp distribution was found for high-oriented and polycrystalline PZT films in their fresh state, while broader distributions with an important reversible component are characteristic of the non-homogeneous and non-oriented ceramics. Particularly, very broad and dispersed FORC distributions were found in ferroelectrics in their fatigued state, due to the local imprint effects [127]. Distributions with multiple maxima were also proposed to describe ferroelectric samples with a certain degree of inhomogeneity, like unpoled ceramics [134]. Thus the FORC distributions are giving fingerprints of the ferroelectric systems on the point of view of their local switching properties. The FORC method was used as additional important tool in the present chapter to investigate the switching properties of BTZ ceramics with various compositions and grain sizes.

(A) Effect of Composition on the P(E) Loops

The minor hysteresis loops obtained for BTZ ceramics with the composition $x = 0.10$ prepared at different sintering temperatures are represented in the figure 29 and ones of the composition $x = 0.15$ in the figure 30. The nested aspect of the minors, the Rayleigh behavior in the sub-switching region and the saturation at high fields of the loops clearly demonstrate the existence of a ferroelectric macroscopic behavior in all these samples. The first-polarization curve as described by the corners of the minors at various field amplitudes has a sharper field-dependence as the sintering temperature is increasing. This is due to the fact that with increasing the grain size, more switchable units switch at the same fields, while for smaller grain sizes, the coercive fields are more distributed.

It can be also observed in figure 29 that the remanent polarization of the BaTi$_{0.9}$Zr$_{0.1}$O$_3$ samples increases from $P_r \cong 1.3$ μC/cm^2 for the sample sintered at $T_s = 1350°C$, to a much higher value of $P_r \cong 6.5$ μC/cm^2, for the sample sintered at $T_s = 1500°C$. It is also noticed that the loops tend to a more rectangular shape with increasing the sintering temperature, thus the switching character is improving when increasing the grain size of the samples. The very high polarization of the sample sintered at $T_s = 1500°C$ and their excellent dielectric properties are also due to the very high density obtained for these samples (Table 2). A remarkable increase of P_r with increasing the annealing temperature has been also reported for polycrystalline thin films of BaTiO$_3$ [135], SrBi$_2$Ta$_2$O$_9$ [136, 137], PbTiO$_3$ [138] and for BaTiO$_3$ ceramics with grain sizes down to 30 nm [111]. The present results fit in the same trend as observed by other authors.

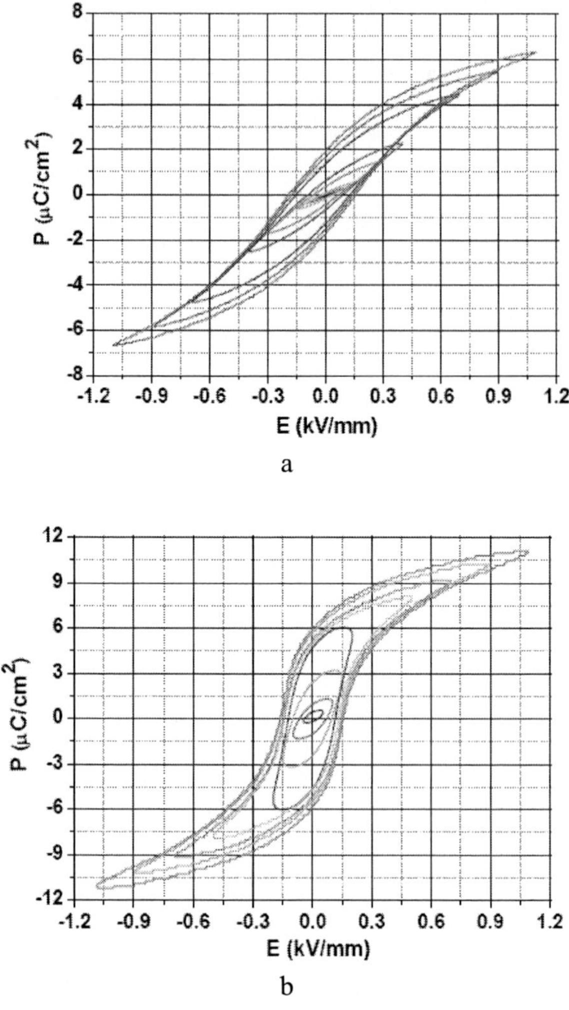

Figure 29. Minor hysteresis loops at a few field amplitudes and $f = 10$ Hz recorded for BaTi$_{0.9}$Zr$_{0.1}$O$_3$ ceramics sintered at temperatures of: (a) 1350°C and (b) 1500°C.

In the figure 30 is presented the minor hysteresis loops for BaTi$_{0.85}$Zr$_{0.15}$O$_3$ ceramic sintered at 1500°C. There is no contribution from extrinsic phenomena and the MHL is well saturated. The coercivities of different dipolar units are distributed starting with low values of the fields and consequently, the MHL is less rectangular than one corresponding to the BaTi$_{0.9}$Zr$_{0.1}$O$_3$ ceramic sintered at the same temperature. The remanent polarization $P_r \cong 4.5$ μC/cm^2 is smaller that of BaTi$_{0.9}$Zr$_{0.1}$O$_3$.

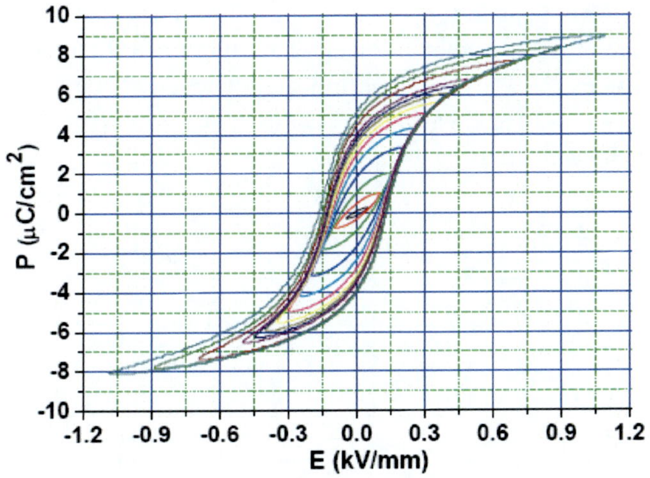

Figure 30. Minor hysteresis loops at a few field amplitudes and f = 10 Hz recorded for BaTi$_{0.85}$Zr$_{0.15}$O$_3$ ceramic sintered at 1500°C.

The influence of composition on the polarization-field response $P(E)$ of BaTi$_{1-x}$Zr$_x$O$_3$ ceramics with x = 0.10 and x = 0.15 sintered at the same temperature (a) or having comparable grain size (b) is presented in the figure 31. The ceramics with lower Zr content has larger saturated hysteresis loop with higher remanent polarization (figure 31 (a). With increasing the Zr content the hysteresis loop became slimmer, indicating that the long range polarization ordering is broken by the presence of heterogeneity in the B-site of the unit cell caused by the Zr addition.

Since the $P(E)$ loops as other functional properties of BZT ceramics are very sensitive to the grain size and density, in the figure 31(b) are compared the same compositions sintered at different temperatures but with similar grain size (1.1 µm for x = 0.10 and 1.8 µm for x = 0.15) and also similar relative densities ($\sim$ 90%). On these two ceramics the effect of Zr addition in changing the switching characteristics of the solid solution is very evident and demonstrate the tendency towards the relaxor state (slimmer $P(E)$ loops, lower coercivity and lower remanent polarization) when the Zr content increases. For comparison sake, the P_r = 6.27 µC/cm^2, E_c = 0.21 kV/mm for x = 0.10 and P_r = 4.5 µC/cm^2, E_c = 0.14 kV/mm for x = 0.15. The sample with x = 0.15 saturates at lower fields than the sample x = 0.10. It results that smaller barriers that separate the bi-stable states of the ferroelectric polarization are induced by higher Zr addition, because at room temperature this composition is closer to its Curie temperature. In addition, closer to the phase transition, the domain walls mobility is higher, allowing the saturation at smaller fields.

In conclusion, the $P(E)$ data confirm the tendency of the system towards the relaxor state with increasing Zr content, as observed also from the dielectric data investigation.

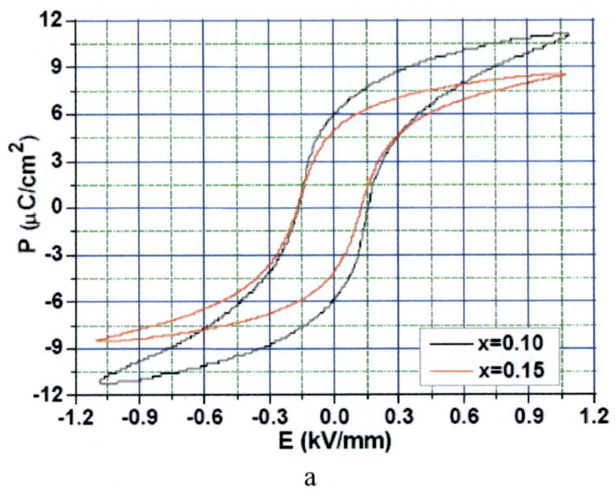

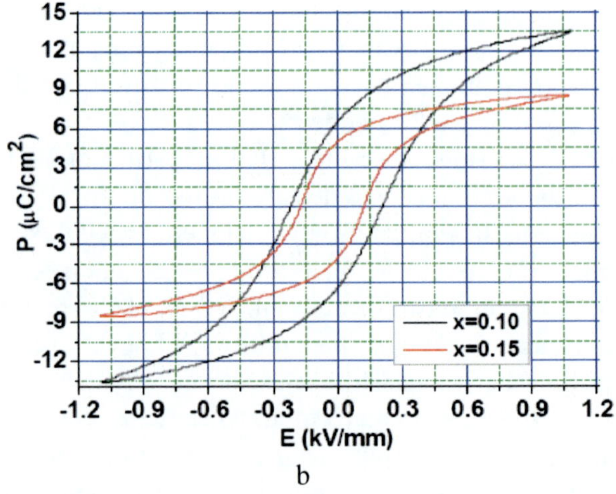

Figure 31. MHL of BaTi$_{1-x}$Zr$_x$O$_3$ ceramics with $x = 0.10$ and $x = 0.15$ at the field amplitude $E_0 = 1.1$ kV/mm and frequency $f = 10$Hz: (a) sintered at the same temperature of 1500°C; (b) having comparable grain size.

(B) Effect of Grain Size on the P(E) Loops

As already mentioned when discussed the minor loops, the grain size has an important effect on the switching properties and consequently, on the characteristics of the $P(E)$ loops. From the SEM analyses it was shown that when the sintering temperature increases from 1350°C to 1500°C, also the grain sizes of the samples BaTi$_{0.9}$Zr$_{0.1}$O$_3$ increase from ~ 0.75 μm to ~ 3.27 μm together with an increase of density (Table 2). It was therefore expected to see differences in the $P(E)$ **loops' shape** as result of this intrinsic grain size effect. Therefore, as already mentioned, the $P(E)$ loops are also strongly dependent on extrinsic factors, particularly on charged and polar defects giving an electrical response overlapped to the ferroelectric intrinsic one. Since these extrinsic phenomena are not known in the series of samples investigated here, we will show in the following just the tendency of the MHL when increasing the grain size.

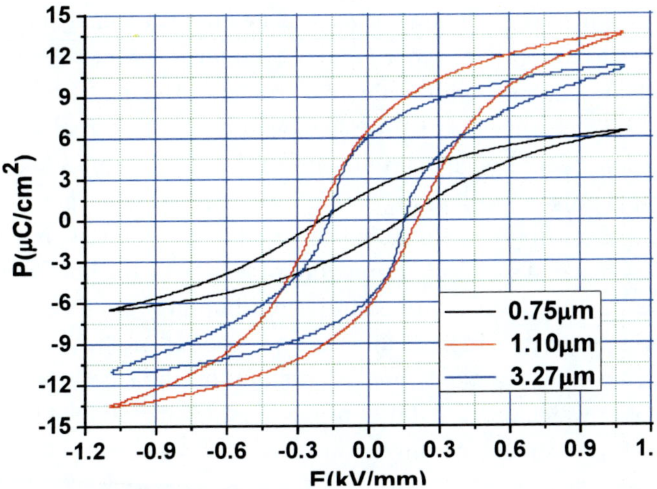

Figure 32. $P(E)$ loops of BaTi$_{0.9}$Zr$_{0.1}$O$_3$ ceramics with different grain size.

As resulted from figure 29, where the minor loops were shown, the sample sintered at lower temperature T_s = 1350°C (grain size of ~ 0.75 μm), presents a very tilted $P(E)$ loop, with a large reversible component, a low rectangularity factor, a small value of the remanent polarization ($P_r \cong$ 2.4 μC/cm^2) and a coercive field of 0.17 kV/mm A higher remanent polarization is obtained for the ceramics sintered at 1400°C and 1500°C, due to their larger grain size. Figure 32 shows that the ceramic with the best density and highest grain size of ~ 3.27 μm sintered at 1500°C has the most rectangular $P(E)$ loop and the lowest coercivity (E_c = 0.15 kV/mm), but its saturation polarization is slightly smaller than for the sample sintered at 1400°C having grain size of ~ 1.1 μm (P_s = 11.12 μC/cm^2), while the remanent polarization is almost the same ($P_r \cong$ 6.5 μC/cm^2). In any case, the MHLs obtained in the present ceramics reflect the tendency found by other authors in systems like BT, BTZ, BST and PZT thin films [137, 139-141].

When comparing the $P(E)$ loops of the different ceramic samples it should be kept in mind the possible differences caused by extrinsic phenomena, as already mentioned before. These extrinsic contribution might be different for the samples sintered at various temperatures and this might explain why a higher saturation polarization was found in the sample with grain size of ~ 1.1 μm and not in the one with higher grain size (~ 3.27 μm), in spite its excellent dielectric properties and almost full densification (99% relative density). Another possible explanation for the non-monotonous reduction of polarization when reducing grain size is related to the anomaly of the dielectric and ferroelectric properties reported in the landmark paper of Arlt *et al.* [140] in the BaTiO$_3$ ceramics with grain size around ~ 1 μm that might take place also in the BTZ ceramics. At a given temperature in the ferroelectric phase, the grain size dependence of the permittivity of BaTiO$_3$ ceramics has a maximum value around this critical grain size of ~1 μm. For this grain size, it was supposed that an optimum twinning of the domain walls within the ceramic grains takes place. As a result, a maximum density of the 90° domain walls is obtained and they give important contributions to the dielectric and ferroelectric properties. In spite there are no reported studies of such effects in BaTiO$_3$-based solid solutions, nor information on the structure and dynamics of the domain walls, the possibility of having a ferroelastic anomaly at a grain size of ~ 1 μm should be further investigated as a possible origin of the observed behavior.

(C) Characterization by FORC Diagrams

The FORC diagram method was employed in this study in order to see changes of the switching character of the BaTi$_{1-x}$Zr$_x$O$_3$ ceramics possible induced by differences of the composition and grain size [94, 95, 142]. The experimental FORCs, required for calculating the FORC diagrams were acquired by cycling the samples between the saturation, E_{sat}, and a variable reversal field, $E_r \in (-E_{sat}, E_{sat})$. The experimental FORCs were recorded under sinusoidal waveform with maximum amplitude of E_0 = 2.5 kV/mm and frequency f = 10 Hz.

Due to the fact that the samples with extreme grain size sintered at 1350°C and 1500°C have well-defined FORCs and a low extrinsic contribution, they are used in the following for a comparative discussion concerning the effect of the grain size on the switching properties, as revealed by the FORC analysis. Therefore, it is important to note that the differences observed in the switching properties of these samples are the results of a combined grain size and porosity effect, since the samples sintered at 1350°C and 1500°C have relative densities of 89.4% and 99%, respectively (Table 2). As noted before, the remanent polarization increases with increasing the sintering temperature, from ~ 3.5 μC/cm^2 to ~ 7 μC/cm^2, whereas the coercivity is almost the same, of ~ 0.25 kV/mm.

The 2D-FORC diagrams and the corresponding 3-D representation of the FORC distributions $\rho(E_r, E)$ are showed in figures 33 (a)-(f). One of the advantages of the FORC method is that the reversible and irreversible contributions to the total polarization can be separated. The irreversible contribution has a maximum located in the fourth quadrant, whereas the reversible component stretches along the first bisecting line in the first and third quadrant. The fields corresponding to the maximum represent the most probable fields for which the highest number of switchable units of the system (causing the highest contribution to the ferroelectric polarization) exists. The reversible component is due to the low-field contributions to the total polarization, *e.g.* the lattice intrinsic contribution and domain wall oscillations. For the mentioned two extreme grain sizes, the reversible part, is higher and extended for a larger range of the fields with both positive and negative polarity in the fine BZT ceramic (figure 33 (a)) by comparison with the coarse one (figure 33 (e)). The irreversible components of the FORC distributions have a well-defined maximum for both samples: a more diffuse one for the fine sample (figure 33 (b)) and a very sharp one for the coarse ceramic (figure 33 (f)).

The FORC distributions of the figure 33 (a) and (e) show a small bias, *i.e.* a shift of the distribution maximum towards negative bias fields, with similar values: E_{bias} = -2.19 kV/mm for the sample sintered at 1350°C and E_{bias} = -2.16 kV/mm for the sample sintered at 1500°C. Being bulk structures with symmetric structures and identical electrodes, the possible origin of this bias as coming from the interfaces ceramics-electrodes is excluded. Since the negative bias is typical to all samples investigated here, it results that this is an intrinsic effect coming either from the processing (one-directional pressing) or by dipolar defects contributions intrinsic to this type of material. The maximum is located at rather low coercivities ($\sim$ 0.67 kV/mm) for both discussed samples. These values indicate low energy barriers for the large majority of irreversible domain walls movements and only a small number of dipolar units are switchable under higher fields. As observed in figure 33 (a) and (e), there is not a separation between the reversible (along the bias axis $E_c = 0$) and the irreversible (for $E_c \neq 0$) components of the polarization on the FORC distribution as in other ferroelectric systems [127, 143], particularly for the fine sample. It results that for this composition, a continuous distribution of energy barriers for switching from zero to non-zero values is characteristic. This means that the system has a mixed ferroelectric − relaxor character, the full relaxor state (superparaelectric state) being ideally characterized by almost zero irreversible FORC component. A similar behavior was reported for some $(Ba,Sr)TiO_3$ compositions [94, 144] and is related to the high degree of local compositional inhomogeneity of the solid solutions, giving rise to broad distributed Curie temperatures and coercivities. Therefore, the sample sintered at 1500°C (figure 33 (e), (f) is much homogeneous from the switching point of view, *i.e.* the largest majority of the dipolar units are switchable at very similar fields, giving rise to a very sharp FORC distribution, almost without reversible component, confirming the excellent ferroelectric properties of this ceramic. This is a result of two effects:

I   larger grains and small non-ferroelectric grain boundary volume resulting in switching characteristics closer to the single-crystal behavior;

II   high density, putting the overall dipolar units into similar boundary conditions.

The enlarging FORC distribution by porosity effects was also observed in PZT ceramics with a high degree of porosity and it was explained by a scattering of the internal and coercive fields as result of different local electrical and mechanical boundary conditions inside the ferroelectric ceramic [145]. This is due to the facts that grains partially surrounded by pores have increased possibility to accommodate the strain resulting from the paraelectric-to-ferrolectric transition and to the fact that gas-solid interfaces have a different energy in comparison to solid-solid interfaces (grain boundaries). In any type of characterization, particularly in the electric ones, there is a serious need to distinguish between intrinsic and extrinsic effects on the macroscopic properties.

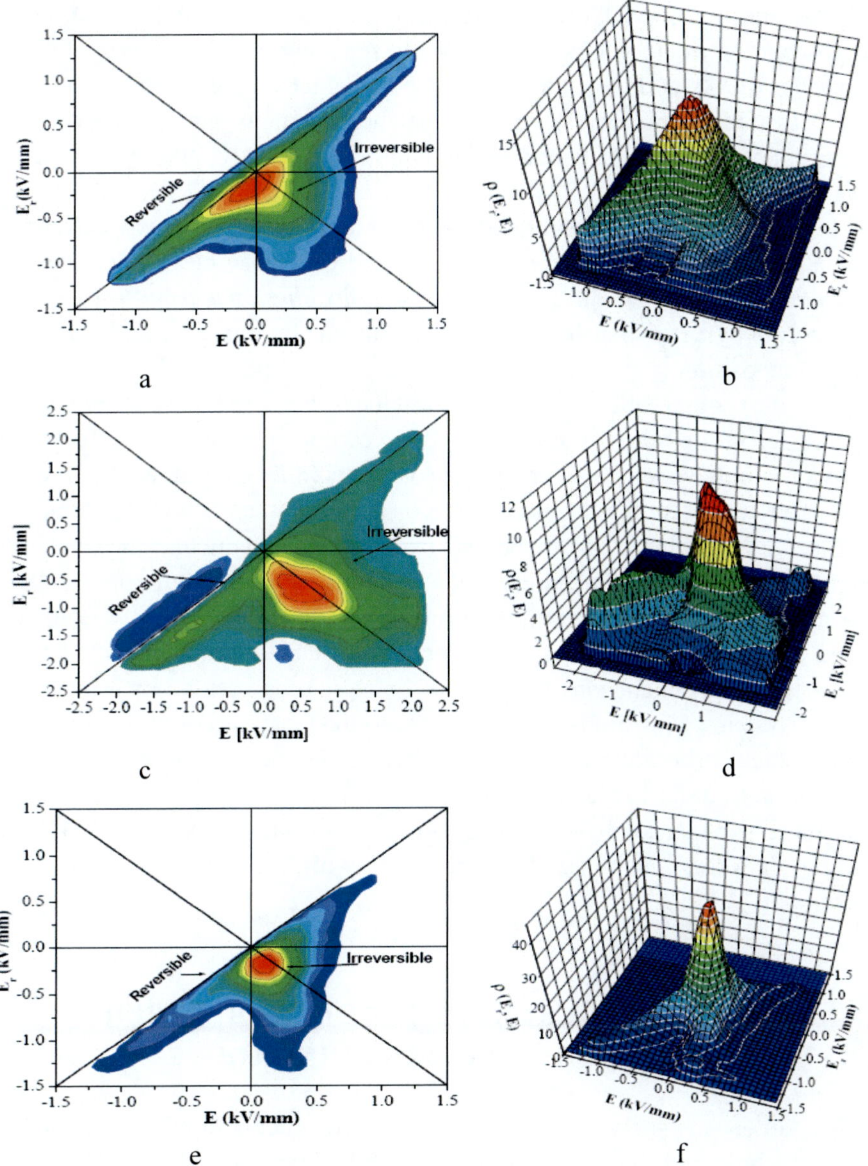

Figure 33. 2D-FORC diagrams and 3D-FORC distributions obtained for BaTi$_{0.9}$Zr$_{0.1}$O$_3$ ceramics, sintered at: (a)-(b)1350°C; (c)-(d) 1450°C and (e)-(f) 1500°C, for f = 10Hz.

Generally, the $P(E)$ hysteresis loop analysis and the FORC investigation presented before proved that all BTZ ceramics investigated here present a macroscopic ferroelectric behavior. With reducing the grain size and by increasing the Zr addition, the ferroelectric character of the solid solution is altered as due to the increasing the local heterogeneity and disorder which disrupt the dipolar long range order and, consequently, the system tends towards its relaxor state.

In conclusion, the corroborated dielectric and hysteresis study performed in BTZ ceramics with a few compositions and grain size clearly demonstrate *the tendency of the ceramic system to change its character from ferroelectric into relaxor* by: *(i)* increasing Zr addition, and *(ii)* by reducing grain size at a given composition.

The last observation allows to conclude that for really producing a superparaelectric (relaxor) state in a BTZ composition with a ferroelectric behavior in bulk state (single crystal or coarse ceramics), a further reduction of grain size down to the nanoscale region is necessary. Further attempt to produce smaller grain sizes below 0.75 μm by reducing the sintering temperature or time produced ceramics with unacceptable high porosity for a relevant electrical investigation. A porosity level of > 5% generally leads to a strong depression of the apparent dielectric constant. Moreover, for a correct comparison and conclusions, the role of grain size have to be investigated in ceramic systems with similar densities, due to the fact that porous ceramics are in completely different electric and mechanical boundary conditions than the dense ones. In order to really obtain dense BTZ ceramics with nanosized grain size, more sophisticated methods of sintering are necessary and much finer starting powders, non-agglomerated and with a narrow particle size distribution.

3.2. Ceramics from Powders Prepared by the Modified Pechini Route

In order to show the influence of the preparation route on the microstructure and functional properties, BTZ ceramics derived from the nanopowders synthesized by the modified Pehini method were also prepared. $Ba(Ti_{1-x}Zr_x)O_3$ (0.1 $\leq x \leq$ 0.2) nanopowders were shaped into pellets and then sintered by alternative methods. The characteristics of the resulted ceramics were compared with those of the ceramic samples prepared by the solid state reaction.

3.2.1 Ceramics Produced by Classical Sintering

(A) Phase Composition and Microstructure

Classical sintering was carried out in air, with a soaking time of 2 hours, at temperatures of 1300° and 1400°C, respectively.

Room-temperature X-ray diffraction data show that, irrespective of the zirconium content, the ceramics are well-crystallized and single phase even after sintering at 1300°C (figure 34)

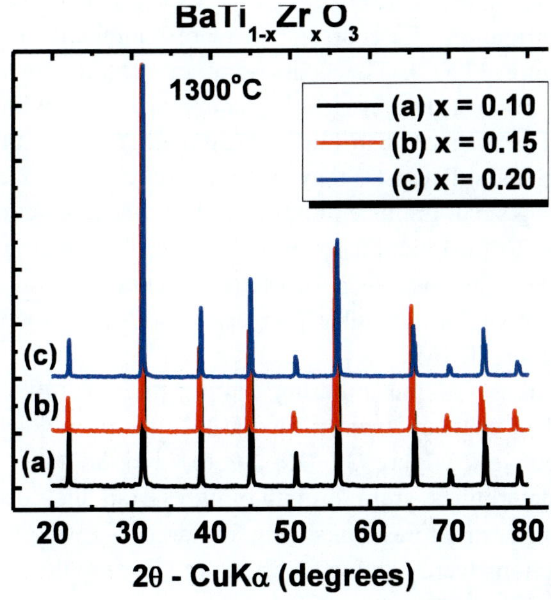

Figure 34. Room temperature XRD pattern of the BaTi$_{1-x}$Zr$_x$O$_3$ ceramics obtained after classical sintering at 1300°C.

For the BaTi$_{0.9}$Zr$_{0.1}$O$_3$ ceramic obtained after classical sintering at 1300°C, a relative dense and homogeneous fine-grained microstructure, with not well defined grain boundaries was pointed out by the corresponding SEM image (figure 35 (a)). As the amount of zirconium in the BTZ solid solution increases (x = 0.15), the proximity of the phase transition induces a microstrure disturbance in comparison with the Zr-poorer ceramic ($x = 0.10$) sintered in the same conditions. Thereby, in this case a bimodal grain size distribution was pointed out (figure 36 (a)). The corresponding SEM image shows that regions with larger (of ~ $15 - 20$ µm), almost polyhedral, non-uniform grains with multilamellar aspect indicating a step-growth mechanism, coexist with extended zones of agglomerated, isotropically interconnected, smaller (with ~ an order of magnitude) grains more homogeneous as shape and size and exhibiting a matrix-like aspect because their not well defined grain boundaries. As a consequence of this bimodal grain distribution, an average grain size value of ~ 9 µm was estimated. Besides, the presence of some inter-granular pores can also be observed, mainly between the adjacent regions of large and small grains (figure 36 (a)). These features indicate discontinuous, abnormal grain growth. With the further increase of the zirconium concentration ($x =$ 0.20), the formation of a coarser, but more homogeneous microstructure was noticed (figure 37 (a)). The coarse-grained regions are more extended and prevail at microstructural level, indicating an obvious tendency to a monomodal grain size distribution. Although even in this case some fine-grained regions still exists, they are rare, much less extended and with the corresponding small grains exhibiting well-defined, spherical shape. The large grains (of ~ 30 µm) exhibit rough aspect, well defined grain boundaries and perfect triple junctions. Because of the accentuated grain growth process, a small amount of intragranular porosity inside the large grains can also be observed (figure 37 (a)).

The increase of the sintering temperature to 1400°C induces obvious morphological changes, mainly for the BTZ compositions with $x = 0.15$ and 0.20, respectively (figure 35 (b), 36 (b) and 37 (b)). Thus the sintering evolution determines grain growth progress, so that denser, homogeneous pore-free microstructures consisting of large (~ $50 - 70$ µm) grains with perfect junctions were obtained, especially for zirconium-richer ceramics ($x =$ 0.15 and 0.20). In this case, the grains relief suggests the evolution of the growth process by the coalescence of smaller grains into larger ones, by the movement of the grain boundaries.

The evolution of the main microstructural features, *i.e.* relative density and average grain size is presented in Table 3.

In comparison with the ceramics obtained by solid state reaction, significantly higher values of the grain average size were estimated from SEM images especially for the Zr-richer ceramic samples ($x = 0.15$ and $x = 0.20$) prepared from nanopowders synthesized by the modified Pechini method and classically sintered at a higher temperature of 1400°C (Table 3).

Table 3. Characteristics of BaTi$_{1-x}$Zr$_x$O$_3$ ceramic samples after sintering

Sample	Annealing temperature of the precursor oxide powder (Pechini) (°C)	Temperature of sintering (°C)	Sintered Density (%)	Average grain size (μm)
BaTi$_{0.9}$Zr$_{0.1}$O$_3$	850	1300	90.50	1.5
		1400	94.33	6.4
BaTi$_{0.85}$Zr$_{0.15}$O$_3$	850	1300	88.80	9.2
		1400	95.37	52.6
BaTi$_{0.8}$Zr$_{0.2}$O$_3$	850	1300	94.82	28.8
		1400	98.72	68.8

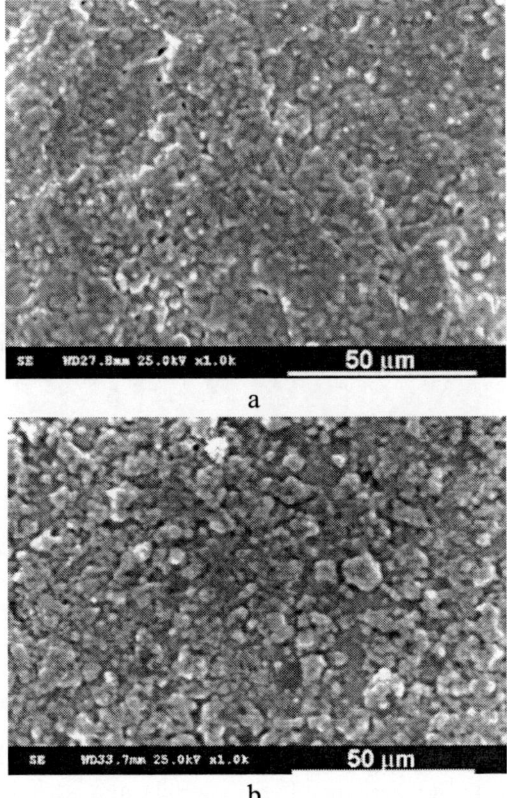

Figure 35. Surface SEM images of BaTi$_{0.90}$Zr$_{0.10}$O$_3$ ceramics prepared from the corresponding nanopowder synthesized by the modified Pechini method and classically sintered at: 1300°C (bar = 50 μm) and (b) 1400°C (bar = 50 μm).

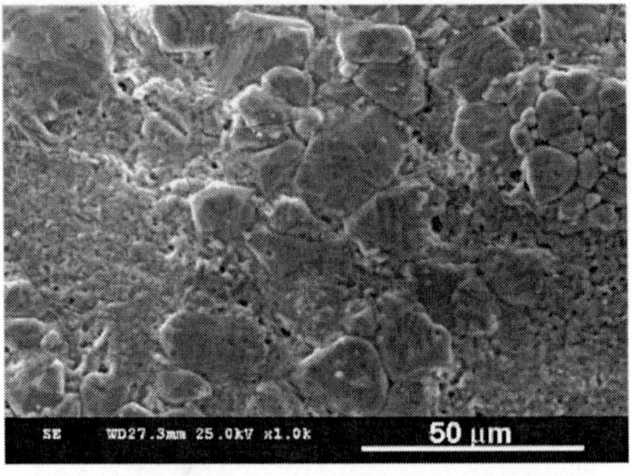

a

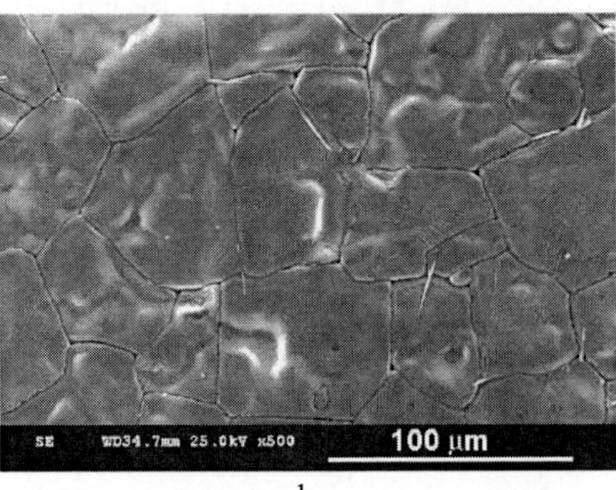

b

Figure 36. Surface SEM images of $BaTi_{0.85}Zr_{0.15}O_3$ ceramics prepared from the corresponding nanopowder synthesized by the modified Pechini method and classically sintered at: (a) 1300°C (bar = 50 μm) and (b) 1400°C (bar = 100 μm).

This microstructural feature demonstrates an obviously higher rate of the grain growth due to the enhanced diffusional processes in the case of the ceramics derived from the powders prepared by the polymer precursor method, than that one corresponding to the ceramics produced by the traditional mixed oxide method. This is probably because in the last case the chemical reaction with the formation of the perovskite lattice slows down the diffusional processes and consequently also the densification, which occurs at higher temperatures, at least for the $BaTi_{0.9}Zr_{0.1}O_3$ (Table 2), in comparison with the first case when the BTZ solid solutions are already formed in the starting oxide powders.

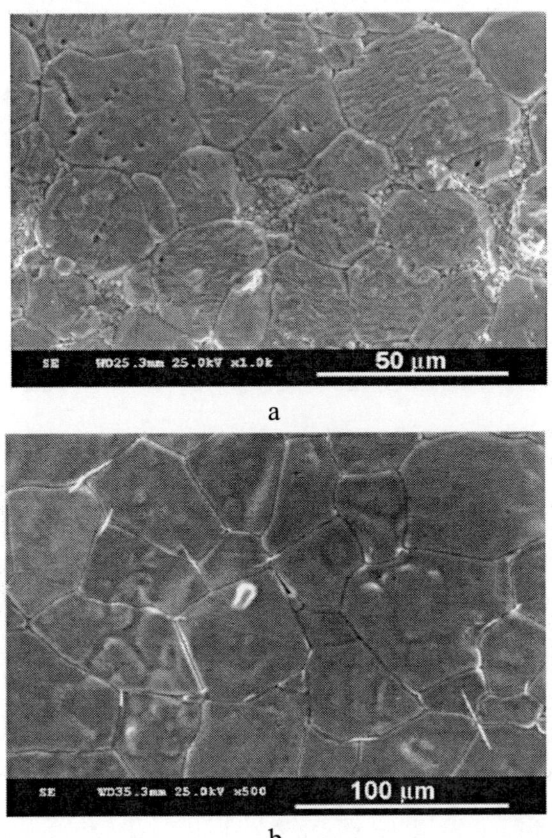

Figure 37. Surface SEM images of BaTi$_{0.80}$Zr$_{0.20}$O$_3$ ceramics prepared from the corresponding nanopowder synthesized by the modified Pechini method and classically sintered at: (a) 1300°C (bar = 50 μm) and (b) 1400°C (bar = 100 μm).

For the ceramics derived from nanopowders prepared by the modified Pechini method, a better densification pointed out by the higher relative density values was achieved. Thus, for a fixed composition *e.g.* BaTi$_{0.9}$Zr$_{0.1}$O$_3$ classically sintered in the same conditions (1400°C/2h), but starting from powders prepared by different methods (*i.e.* mixed oxide method and modified Pechini procedure) a higher relative density (of 94.33%) was obtained for the ceramic prepared by the polymer precursor method, than that one of 90.70% (Table 2) estimated for the sample processed by the traditional ceramic route.

Moreover, in the case of compositions with higher Zr content ($x = 0.15$ and 0.20) the relative density reaches higher values for ceramics prepared via Pechini method and classically sintered at 1400°C than that of ceramics prepared by mixed oxide route and sintered even at a higher temperature of 1500°C, as the same plateau was maintained.

The still low grain size, as well as the lower density values for the Zr-richer ceramics obtained by the solid state method after sintering at the high temperature of 1500°C can be explained in terms of the disturbance induced in necks growth, affecting the grain boundaries movement by formation of a Kirkendall-type porosity, which is well-known to hamper densification in such heterogeneous systems as $BaTiO_3$-based solid solutions with higher solute concentrations [63, 146]. One can conclude that the lower grain growth rate and the formation of intergranular pores originate in the difference in diffusion rates between the two metallic species involved in the mixed crystals formation.

Concerning the ceramics preparing by the modified Pechini route, only for the $BaT_{0.85}Zr_{0.15}O_3$ sample sintered at 1300°C the relative density was slightly lower than 90%, most likely because of a little bit higher intergrain porosity originated in this case in the microstructural disturbance induced by the discontinuous (abnormal) grain growth.

(B) Dielectric Properties

The preliminary dielectric characterization performed at room temperature (figures 38 and 39) shows only a slight frequency dependence of the dielectric permittivity over the all frequency range, irrespective of the sintering temperature (figure 38 (a) and figure 39 (a)). A higher frequency dispersion is obvious for the dielectric loss, mainly in the low frequency range ($f = 1 - 100$ Hz) (figure 38 (b) and figure 39 (b)). At higher frequencies above 1 kHz, all the present ceramics are good dielectrics, with losses tan $\delta < 4\%$. As shown in the Figs. 38 (a) and 39 (a), an obvious composition dependence of the dielectric permittivity at a fixed frequency is noticed. For the ceramics sintered at 1300°C, the maximum permittivity (in the range of $9336 - 11204$) is achieved by the composition $x = 0.20$. The ceramic sample with the same composition, but sintered at 1400°C, presents even higher maximum permittivity values of $13030 - 14514$ in the range of frequencies of $1 - 10^5$ Hz, probably due to the higher densification and crystallinity. The high values of the dielectric permittivity might indicate the proximity of the ferroelectric – paraelectric phase transition at room temperature for the ceramic with the nominal $BaTi_{0.8}Zr_{0.2}O_3$ composition.

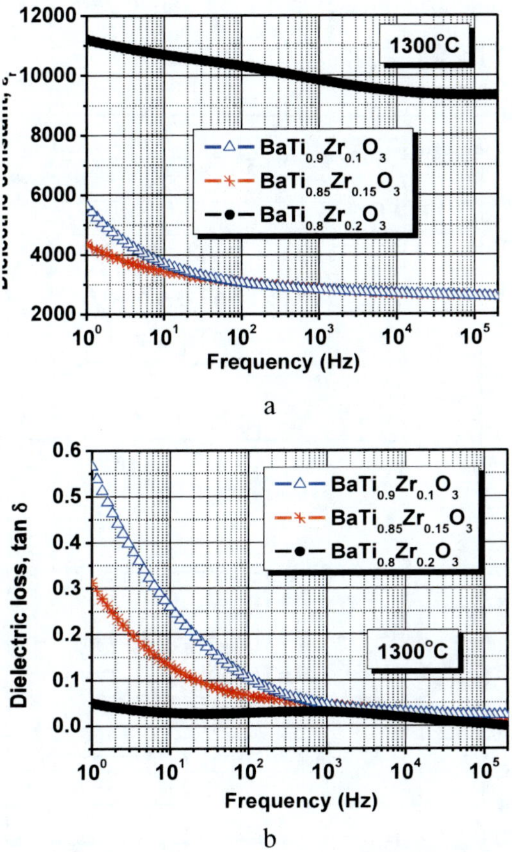

Figure 38. Room temperature frequency dependence of the dielectric properties for BaTi$_{1-x}$Zr$_x$O$_3$ ceramics prepared from the corresponding nanopowders synthesized by the modified Pechini method and classically sintered at 1300°C: (a) dielectric constant; (b) dielectric loss.

By comparison with ceramics prepared by solid state presented before, much higher permittivity and low losses are found for ceramics produced by Pechini powders. The ceramic with composition $x = 0.20$ sintered at 1400°C shows excellent dielectric properties, with a very high permittivity and losses below 2% at room temperature, being among the best ever reported values in BTZ ceramics [10-13, 34-37, 40-43] and these properties are related to optimum microstructural characteristics. Further investigations of the dielectric and ferroelectric properties at various temperatures including the ferroelectric – paraelectric phase transition range will clarify the specific behavior of these ceramics.

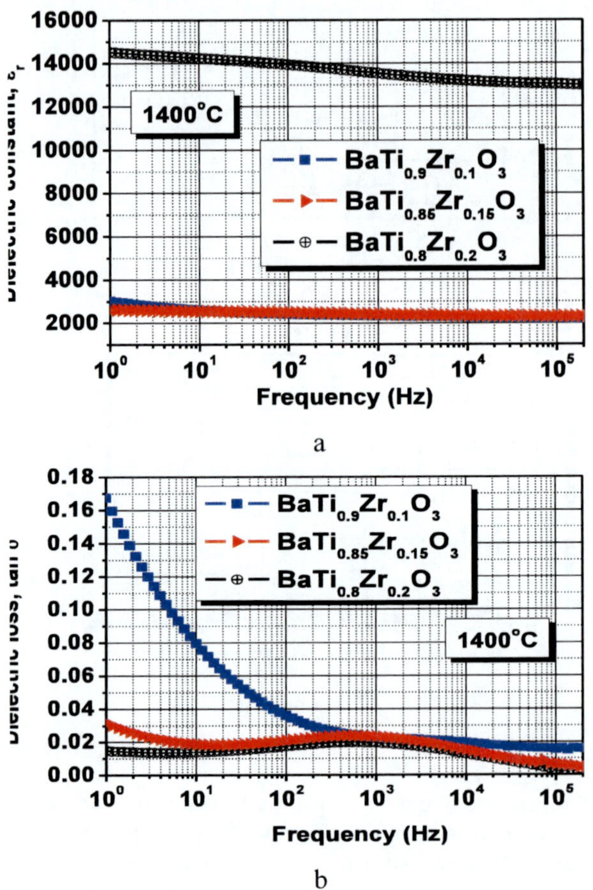

Figure 39. Room temperature frequency dependence of the dielectric properties for BaTi$_{1-x}$Zr$_x$O$_3$ ceramics prepared from the corresponding nanopowder synthesized by the modified Pechini method and classically sintered at 1400°C: (a) dielectric constant; (b) dielectric loss.

3.2.2. Ceramics Produced by Spark Plasma Sintering (SPS)

Dense ceramics with grains below 100 nm have been obtained only in recent years, particularly for pure BaTiO$_3$, by using special densification methods, like ultra-high pressure sintering in a multianvil press (grain size down to 70 nm, 98% relative density) [147], combined sintering method (grain size down to 90 nm, 99% relative density) [148] and most recently by using Spark Plasma Sintering (SPS) technique [149]. The last method is actually a hot-press process which combines the pressure and the plasma generator. In the last years, spark plasma sintering was quite often used as an alternative method to the classical sintering procedure in order to obtain highly densified, nanostructured ceramics from powders prepared by various wet chemical methods [150-159]. Thereby, BaTiO$_3$ ceramic bodies with a minimum grain growth and, therefore, exhibiting grain size (GS) down to 30 nm and relative density of 95 – 99% were produced by this procedure. [109-111, 160].

Unlike the pure BaTiO$_3$, spark plasma sintered ceramics derived from homo- or aliovalent barium titanate solid solutions are much less studied from the point of view of their structural, microstructural and electrical characteristics [161]. No publications on the size effects down to nanoscale in BTZ solid solutions are known. Having in mind the characteristics of the SPS technique, ultrafine BTZ powders prepared by the modified Pechini route were recently chosen to be densified by this method, in order to obtain fine ceramics with grain size below 100 nm. Due to the relatively low temperatures and short sintering time, grain growth is largely inhibited, while densification proceeds at a high rate.

For this reason, a first attempt to obtain and characterize spark plasma sintered BTZ ceramics was performed. A few results concerning such ultrafine ceramics are presented in the following. It is worthy to mention that this is a complex study still under way and only partial results can be for the moment presented.

Previous papers on SPS barium titanate ceramics pointed out that the grain growth rate during this type of sintering is very sensitive to the type of the wet-chemical method used for the powder preparation [150, 151]. Thus, for certain methods (as the sol-gel acetate, methanol method or sol-precipitation), in spite of relative small starting particles size, their growth is very fast with temperature increase. Therefore, the BaTi$_{0.85}$Zr$_{0.15}$O$_3$ nanopowder synthesized by the modified Pechini method was considered appropriate for this purpose. An amount of ~ 5 g of the precursor oxide powder was put into a graphite mold. Spark plasma sintering was carried out in vacuum, at 1000°C for 2 min, with a heating rate of 100°C/min and under a pressure of 76 MPa.

(A) Phase Composition, Structure and Microstructure

Room-temperature X-ray diffraction data show that the spark plasma sintered BaTi$_{0.85}$Zr$_{0.15}$O$_3$ ceramic is also well-crystallized and single phased (figure 40). From structural point of view, it exhibits the same cubic symmetry of the unit cell as the corresponding precursor nanopowder, having a lattice parameter of 4.0299(12) Å, a unit cell volume of 65.44(6) Å, a crystallite average size of 709(57) Å and internal strains of $0.35(5) \times 10^{-3}$. This cubic symmetry of the unit cell indicates the lack of the ferroelectric long range order (LRO). Knowing that lattice parameters calculation from XRD data involves an average over at least 10,000 unit cells, it results that this investigation method is not effective to estimate a local order in such of nanosystems. Therefore, further complementary analyses as Raman spectroscopy are required to establish if the unit cell symmetry is the normal cubic one specific to a real paraelectric state, or it has to speak in terms of the so-called "pseudocubic" form originated either from the complex distortions in the perovskite lattice because of the high internal strains induced by the small size effect (superparaelectric state), or to the presence of an only short range ordering (SRO) of small polar nanoregions (PNR) created by nanoscaled chemical nonhomogeneity, involving a relaxor-like state.

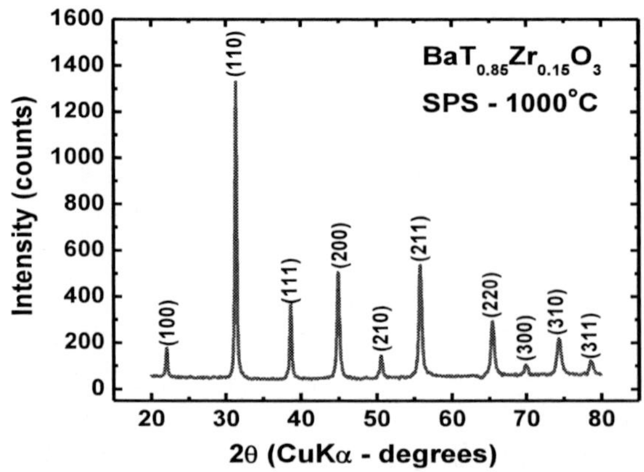

Figure 40. Room temperature XRD pattern of the SPS sintered $BaTi_{0.85}Zr_{0.15}O_3$ ceramic (at 1000°C / 2 min under a pressure of 76 MPa).

SEM-FEG analysis performed on the fracture of the sample indicated that a dense (relative density of ~ 97%), microstructurally homogeneous ceramic was obtained by spark plasma sintering in the mentioned conditions (figure 41 (a)). The higher magnification image (figure 41(b)) shows a grain average size of ~ 75 nm, very close to the value of the average particle size of the precursor oxide powder synthesized by the modified Pechini method. This means that unlike the classical thermal treatment, in this non-conventional sintering process grain growth is strongly inhibited, so that the particle size is almost preserved in the fine-grained, nanocrystalline ceramic.

(B) Dielectric Behavior

For the spark plasma sintered BTZ ceramic sample, in all the frequency range investigated here and at room temperature, significantly lower values of the dielectric permittivity (mainly at higher frequency, $f > 10^3$ Hz) and higher dielectric losses (with ~ an order of magnitude) than those corresponding to the classically sintered ceramic of the same composition and prepared from the same nanopowder were recorded (figure 42 (a) and (b)). For example, at a fixed frequency of 1 kHz, a classically sintered (at 1300°C) BaTi$_{0.85}$Zr$_{0.15}$O$_3$ ceramic exhibits a dielectric permittivity of 9852 and tan $\delta = 0.03$, whereas the SPS ceramic reaches a permittivity value of 531 and tan $\delta \sim 0.3$.

The frequency dependence of the dielectric constant also differs, ε_r showing a drastic drop at lower frequency (bellow 100 Hz) for the spark plasma sintered ceramic (figure 42). The reduction of permittivity at very low frequencies is related to extrinsic dc conductivity which in the case of SPS ceramics is related most probably with insufficient re-oxidation during cooling. The SPS process is very fast, it takes place in reducing atmosphere and consequently, the resulted ceramics are normally semiconductors due to the insufficient re-oxidation. Therefore, a supplementary post-sintering thermal treatment (annealing) at appropriate temperature is required for the complete re-oxidation.

The oxygen vacancies are one of the most common defects present in perovskites, caused by the high firing temperatures and/or low partial Oxygen pressures and this effect is particularly strong in SPS ceramics to the specific processing conditions [162, 163]. The contribution of oxygen vacancy to the dielectric relaxation can be three-fold:

I   It can hop in the lattice, leading to the associated dipolar moment reorientation [164];

II   It is creating conducting electrons through the ionizations: $V_O \Leftrightarrow V_O^{\bullet} + e'$, $V_O^{\bullet} \Leftrightarrow V_O^{\bullet\bullet} + e'$; the conducting electrons in their turn can hop from one localized state to another one in the lattice, equivalent to the reorientation of effective electric dipoles [165, 166];

III Different levels of oxygen deficiency in different spatial regions within the sample can lead to electric heterogeneity even in single phase materials, leading to interfacial polarization (*Maxwell-Wagner* or space charge relaxation mechanism) [118, 167, 168]. The grain boundary in polycrystalline ceramics, various defect structures (such as twin boundaries, dislocations) in single crystals and interphases in composites play an important role in this process. All the mentioned mechanisms are thermally activated processes and are rather difficult to be separated as effects from the dielectric characterisation.

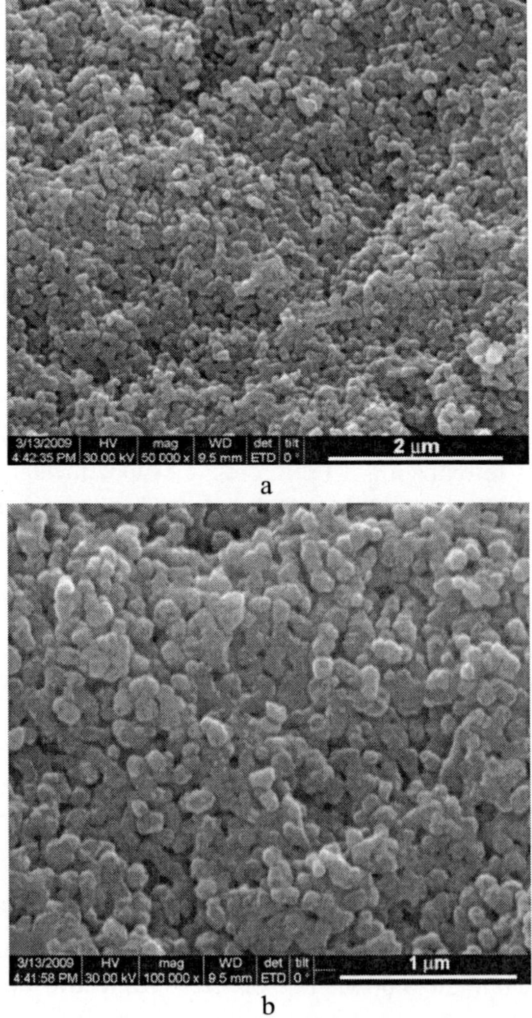

Figure 41. SEM images on fracture of the spark plasma sintered $BTi_{0.85}Zr_{0.15}O_3$ ceramic: (a) general view (bar = 2 μm) and (b) detail (bar = 1 μm).

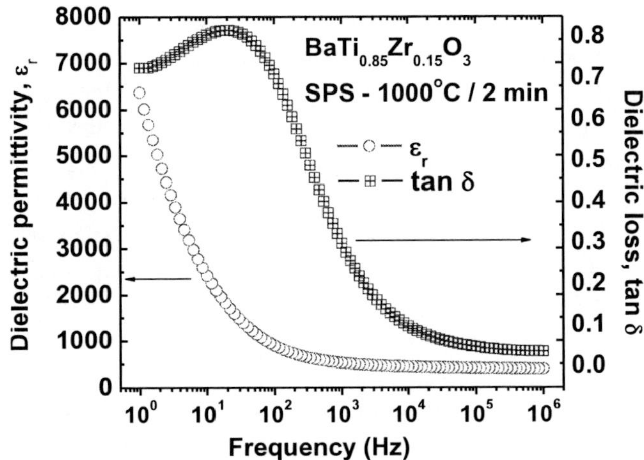

Figure 42. Frequency dependence of the dielectric properties for BTZ15 ceramic obtained after spark plasma sintering at 1000°C / 2 min.

Beside the relaxation at very low frequencies, there is another phenomenon noticed in the evolution of the dielectric loss, which is shifted from kHz toward lower frequency values (~ 1 Hz, see the maximum of tan δ in the figure 42) in comparison with the classical sintered sample (figure 38 (b) and 39 (b)). If this relaxation phenomenon is also related to the oxygen vacancies, we can admit that in the nanostructured ceramic the relaxation process is much slower than in the normal ceramic as result of the large number of interfaces by comparison with the coarse ceramic. Defect charges (including oxygen vacancies) are trapped at interface levels and higher energy or longer time is necessary to produce a polarization contribution than in bulk ceramics. Further dielectric investigations versus temperature are required in order to understand the dielectric relaxation process giving extrinsic contributions to the dielectric and ferroelectric properties in such ceramics.

It is worthy to mention that also *the intrinsic component to the dielectric permittivity*, which can be normally considered as being the response at high frequencies, *is strongly grain size dependent.*

A strongly depressed permittivity is found in the SPS ceramics as result of its nanostructured character (grain size down to 75 nm): ε_r ~ 395 and tan δ ~ 3.5% at 1MHz in the SPS ceramic, by comparison with ε_r ~ 2300 and tan δ ~ 0.46% for the ceramic with $x = 0.15$ sintered at 1400°C, at the same frequency. This behavior was observed for BaTiO₃ nanoceramics [109-111, 160] and is one of the typical effects resulted from the strong reduction of grain size at nanoscale. It is related to the intrinsic loss of ferroelectricity in submicron sized $BaTiO_3$ related to the increased atomic positional disorder in spatially confined physical systems which are disturbing the ferroelectric long range order [169]. Another effect causing the strong observed depression of the dielectric constant is related to the role of a low-permittivity grain boundary layer (so-called *dead layer*). According to these observations, to describe the dielectric properties of the dense ultrafine BTZ ceramics, it seems adequate to adopt a biphasic model in which the ceramic grain is composed by a high permittivity core (which keeps its properties down to this scale) and a low-permittivity non-ferroelectric grain boundary layer, as suggested by "effective field" models, causing a "dilution effect" on permittivity [170].

A detailed analysis of the properties of SPS ceramics in relation with their grain size is further necessary, in order to understand the intrinsic/extrinsic interplay effects on the dielectric constant and polarization, the grain boundary phenomena, the role of oxygen vacancies and the tendency towards a potential relaxor state, which normally would be favored by the grain size reduction.

4. CONCLUSIONS

Wet-chemically prepared Ba(Ti,Zr)O$_3$ powders exhibit various characteristics as structural parameters, particle size, agglomeration tendency and morphology depending on the type of the processing route. The various starting materials and procedures reported for the main wet-chemical techniques could explain the divergent results obtained sometimes in the case of small zirconium-doped barium titanate particles and the controversial mechanisms concerning the transformation of precursors to the final Ba(Ti,Zr)O$_3$ powders. Thus, a lower annealing temperature of 850°C was required to obtain single-phase and well-crystallized Ba(Ti$_{0.85}$Zr$_{0.15}$)O$_3$ nanopowders by the mixed molecular precursor prepared via oxalate route, as well as from the polymeric precursor synthesized by the modified Pechini method, than the powder of the same composition prepared by sol-precipitation, which remains di-phasic (BaTiO$_3$ and BaZrO$_3$) even after annealing at 1100°C.

Regarding, the Ba(Ti,Zr)O$_3$ ceramics their functional properties strongly depends not only on the composition (*i.e.* zirconium content), but also on the microstructure (grain size, porosity, microstructural homogeneity) induced by the preparation route, sintering type and processing parameters.

The size effects in fine-grained ceramics prepared by solid state reaction favor the ferroelectric − relaxor crossover, which occurs at lower zirconium content than in the coarse-grained Ba(Ti,Zr)O$_3$ ceramics.

The characteristics of the samples processed by the traditional ceramic method were compared with those obtained for the classically sintered ceramics prepared from the nanopowders synthesized by the modified Pechini method. As expected, higher densification and consequently better dielectric properties (higher dielectric constant and lower dielectric losses) were obtained after classically sintering at lower temperatures for the ceramics prepared by the modified Pechini method than those of the ceramics with similar composition produced by the traditional mixed oxide route.

On the other hand, the effect of the sintering procedure on the microstructural and dielectric behavior of $BaTi_{0.85}Zr_{0.15}O_3$ ceramics prepared by modified the Pechini method was studied, by an attempt to obtain ultrafine-grained ceramics with grain size down to 100 nm by spark plasma sintering. The size effects in such nanostructured ceramics are expected to induce a relaxor behavior. Therefore, further investigations are required in order to elucidate the structural features and the temperature dependence of the dielectric properties for these non-conventionally spark plasma sintered BTZ ceramics.

ACKNOWLEDGEMENTS

The contributions of Dr. Cristina E. Ciomaga to the study of BTZ ceramics prepared by solid state reaction, of Dr. Daniela Cristina Berger for the preparation of the BTZ ceramics by the modified Pechini method, of Dr. Gilbert Fantozzi for the assistance in the SPS sintering and of dr. Laurentiu Stoleriu for the calculations of FORC diagrams are highly acknowledged.

REFERENCES

[1] A. Rae, M. Chu, V. Ganine, "Barium titanate – Past, present and future", pp. 1–12 in *Ceram. Trans. 2007, vol. 100, Dielectric Ceramic Materials*, Ed. by K.M. Nair, A.S. Bhalla, The American Ceramic Society, Westerville, OH, 1999.

[2] X. Ren, "Large electric-field-induced strain in ferroelectric crystals by point-defect-mediated reversible domain switching", *Nat. Mater.*, 3 (2004) 91-94.

[3] M.E. Lines, A.M. Glass, *Principles and applications of ferroelectrics and related materials*, Clarendon, Oxford, 1977.

[4] G.H. Haertling, „Ferroelectric Ceramics: History and Technology", *J. Am. Ceram. Soc.*, 82 (1999) 797-818.

[5] J.M. Wilson, "Barium titanate", *Am. Ceram. Soc. Bull.*, 74 (1995) 106-110.

[6] J.S. Capurso, A.A. Bologna, W.A. Schulze, "Processing of laminated $BaTiO_3$ structures for stress-sensing applications", *J. Am. Ceram. Soc.*, 78 (1995) 2476-2480.

[7] T.G. Reynod III, "Application Space Influences. Electronic *Ceramic* Materials", *Am. Ceram. Soc. Bull.*, 80 (2001) 29-33.

[8] C.A. Randall, "Scientific and Engineering Issues of the State-of-the-Art and Future Multilayer Capacitors", *J. Ceram. Soc. Japan*, 109 (2001) S2-S6.

[9] K. Uchino, *Ferroelectric devices*, Marcel Dekker, Inc., New York, 2000.

[10] J. Ravez, A. Simon, "Temperature and frequency dielectric response of ferroelectric ceramics with composition $Ba(Ti_{1-x}Zr_x)O_3$", *J. Eur. Solid State Inorg. Chem.*, 34 (1997) 1199-1209.

[11] L. Mitoşeriu, C.E. Ciomaga, I. Dumitru, L.P.Curecheriu, F. Prihor, A. Guzu, "Study of the frequency-dependence of the complex permittivity in Ba(Zr,Ti)O$_3$ ceramics: evidences of the grain boundary phenomena", *J. Optoelectron. Adv. Mater.*, 10 (2008) 1843-1846.

[12] C.E. Ciomaga, M. Viviani, M.T. Buscaglia, V. Buscaglia, R.V. Calderone, L.Mitoşeriu, A. Stancu, P. Nanni, "Preparation and characterisation of the Ba(Zr,Ti)O$_3$ ceramics with relaxor properties", *J. Optoelectron. Adv. Mater.*, 27 (2007) 4061-4064.

[13] C.E. Ciomaga, R. Calderone, M.T. Buscaglia, M. Viviani, V. Buscaglia, L. Mitoşeriu, A. Stancu, P. Nanni, "Relaxor properties of Ba(Zr,Ti)O$_3$ Ceramics", *J. Optoelectron. Adv. Mater.*, 8 (2006) 944-948.

[14] V Kumar1, I Packia Selvam, K Jithesh, P V Divya, „Preparation and dielectric characteristics of nanocrystalline Ba(M$_x$Ti$_{1-x}$)O$_3$", *J. Phys. D: Appl. Phys.*, 40 (2007) 2936-2940.

[15] V. Müller, H. Beige, H.-P. Abicht, „Non-Debye dielectric dispersion of barium titanate stannate in the relaxor", and diffuse phase-transition state", *Appl. Phys. Lett.*, 84 (2004) 1341-1343.

[16] V. V. Shvartsman, W. Kleemann, J. Dec, Z. K. Xu, S. G. Lu, „Diffuse phase transition in BaTi$_{1-x}$Sn$_x$O$_3$ ceramics: An intermediate state between ferroelectric and relaxor behavior", *J. Appl. Phys.*, 99 (2006) 124111.

[17] X. Wei, X. Yao, "Preparation, structure and dielectric property of barium stannate titanate ceramics", *Mater. Sci. Eng. B*, 137 (2007) 184-188.

[18] X. Wei, Y. Feng, X. Wan, X. Yao, "Evolvement of dielectric relaxation of barium stannate titanate ceramics", *Ceram. Int.*, 30 (2004) 1397-1400.

[19] X. Wei, Y. Feng, X. Wan, X. Yao, "Dielectric properties of barium stannate titanate ceramics under bias field", *Ceram. Int.*, 30 (2004) 1401-1404.

[20] S. Marković, D. Usković, "The master sintering curves for BaTi$_{0.975}$Sn$_{0.025}$O$_3$/ BaTi$_{0.85}$Sn$_{0.15}$O$_3$ functionally graded materials", *J. Eur. Ceram. Soc.*, 29 (2009) 2309-2316.

[21] H.Y. Tian, Y. Wang, J. Miao, H.L.W. Chan, C.L. Choy, „Preparation and characterization of hafnium doped barium titanate ceramics", *J. Alloy Compd.*, 431 (2007) 197-202.

[22] S. Anwar, P R Sagdeo, N P Lalla, "Crossover from classical to relaxor ferroelectrics in BaTi$_{1-x}$Hf$_x$O$_3$ ceramics", *J. Phys.: Condens. Matter.*, 18 (2006) 3455-3468.

[23] S. Anwar, P R Sagdeo, N P Lalla, „Ferroelectric relaxor behavior in hafnium doped barium-titanate ceramic", *Solid State Commun.*, 138 (2006) 331-336.

[24] S. Anwar, P R Sagdeo, N P Lalla, „Study of the relaxor behavior in $BaTi_{1-x}Hf_xO_3$ $(0.20 \leq x \leq 0.30)$ ceramics", *Solid State Sci.*, 9 (2007) 1054-1060.

[25] S. Anwar, P R Sagdeo, N P Lalla, „Locating the normal to relaxor phase boundary in $Ba(Ti_{1-x}Hf_x)O_3$ ceramics", *Mater. Res. Bull.*, 43 (2008) 1761-1769.

[26] D. Makovec, Z. Samardžija, D. Kolar, „Solid Solubility of Cerium in $BaTiO_3$" *J. Solid State Chem.* 123 (1996) 30-38.

[27] A. Chen, Y. Zhi, J. Zhi, P.M. Vilarinho, J.L. Baptista, "Synthesis and Characterization of Ba $(Ti_{1-x}Ce_x)O_3$ Ceramics", *J. Eur. Ceram. Soc.*, 17 (1997) 1217-1221.

[28] D. K. Hennings, B. Schreinemacher, H. Schreinemacher, „High Permittivity Ceramics with High Endurance", *J. Eur. Ceram. Soc.*, 13 (1994) 81-88.

[29] G. H. Jonker, W. Kwestroo, "The ternary System $BaO-TiO_2-SnO_2$ and $BaO-TiO_2-ZrO_2$", *J. Am. Ceram. Soc.*, 41 (1958) 390-394.

[30] T. N. Verbitskaya, G. S. Zhdanov, Yu. N. Venevtsev, and S. P. Solov'ev, "Electrical and X-Ray diffraction studies of the $BaTiO_3-BaZrO_3$ system" *Sov. Phys. - Crystallogr. (Engl. Transl.)*, 3 (1958) 182-192.

[31] R. C. Kell, N. J. Hellicar, "Structural Transitions in Barium Titanate-.Zirconate Transducer Materials", *Acustica*, 6 (1956) 235-238.

[32] E. J. Brajer, Polycrystalline Ceramic Materials. U.S. Pat, 1955, No. 2 708 243.

[33] F. Kulscar, Fired Ceramic Barium Titanate Body. U.S. Pat, 1956, No. 2 735 024.

[34] D. Hennings, A. Schnell, G. Simon "Diffuse Ferroelectric Phase Transitions in $Ba(Ti_{1-y}Zr_y)O_3$ Ceramics", *J. Am. Ceram. Soc.*, 65 (1982) 539-544.

[35] Z. Yu, C. Ang, R. Guo, and A. S. Bhalla, "Piezoelectric and strain properties of $Ba(Ti_{1-x}Zr_x)O_3$ ceramics", *J. Appl. Phys.*, 92 (2002) 1489-1493.

[36] *A. Simon, J. Ravez*, M. Maglione, "The crossover from a ferroelectric to a relaxor state in lead-free solid solutions", *J. Phys. Condens. Matter.*, 16 (2004) 963-970.

[37] X.G. Tang, K.H. Chew, H.L.W. Chan, "Diffuse Phase Transition and Dielectric Tunabilibty of $Ba(Zr_yTi_{1-y})O_3$ Relaxor Ferroelectric Ceramics", *Acta Mater.* 52 (2004) 5177-5183.

[38] L. P. Curecheriu, R. Frunză, A. Ianculescu, "Dielectric properties of the $BaTi_{0.85}Zr_{0.15}O_3$ ceramics prepared by different techniques", *Proc. Appl. Ceram.*, 2 (2008) 81-88.

[39] F. M. Tufescu, L. Curecheriu, A. Ianculescu, C.E. Ciomaga, L. Mitoşeriu, "High-voltage tunability measurements of the $BaZr_xTi_{1-x}O_3$ ferroelectric ceramics", *J. Optoelectron. Adv. Mater.*, 10 (2008) 1894-1897.

[40] L. P. Curecheriu, F. M. Tufescu, A. Ianculescu, C. E. Ciomaga, L. Mitoşeriu, A. Stancu, "Tunability characteristics of $BaTiO_3$-based ceramics: Modeling and experimental study", *J. Optoelectron. Adv. Mater.*, 10 (2008) 1792-1795.

[41] Z. Yu, A. Chen, R. Guo, A. S. Bhalla, "Dielectric properties and high tunability of $Ba(Ti_{0.7}Zr_{0.3})O_3$ ceramics under dc electric field", *Appl. Phys. Lett.*, 81 (2002) 1285-1287.

[42] J. Zhai, X. Yao, L. Zhang, B. Shen, "Dielectric nonlinear characteristics of $Ba(Zr_{0.35}Ti_{0.65})O_3$ thin films grown by a sol-gel process", *Appl. Phys. Lett.*, 84 (2004) 3136-3138.

[43] X.G. Tang, K.H. Chew, H.L.W. Chan, "Effect of Grain Size on the Dielectric Properties and Tunabilies of Sol-Gel Derived $Ba(Zr_{0.2}Ti_{0.8})O_3$ Ceramics", *Solid State Commun.*, 131 (2004) 163-168.

[44] Ph. Sciau, G. Calvarin.G, J. Ravez., "X-ray diffraction study of $BaTi_{0.65}Zr_{0.35}O_3$ and $Ba_{0.92}Ca_{0.08}Ti_{0.75}Zr_{0.25}O_3$ compositions: influence of electric field", *Solid State Commun.*, 113 (1999) 77-82.

[45] J. Kreisel, P. Bouvier, M. Maglione, B. Dkhil, A. Simon, "High-pressure Raman investigation of the Pb-free relaxor $BaTi_{0.65}Zr_{0.35}O_3$", *Phys. Rev. B*, 69 (2004) 092104.

[46] M. Nagasawa, H. Kawaji, T. Tojo, T. Atake, "Absence of the heat capacity anomaly in the Pb-free relaxor $BaTi_{0.65}Zr_{0.35}O_3$", *Phys. Rev. B*, 74 (2006) 132101.

[47] S. Ke, H. Fan, H. Huang, H. L. W. Chan, S. Yu, "Dielectric dispersion behavior of $Ba(Zr_xTi_{1-x})O_3$ solid solutions with a quasiferroelectric state", *J. Appl. Phys.*, 104 (2008) 034108.

[48] A.A. Bokov, M. Maglione, A. Simon, Z.-G. Ye, "Dielectric Behavior of $Ba(Ti_{1-x}Zr_x)O_3$ Solid Solution", *Ferroelectrics*, 337 (2006) 169-178.

[49] A.A. Bokov, M. Maglione, Z.-G. Ye, "Quasi-ferroelectric state in Ba(Ti$_{1-x}$Zr$_x$)O$_3$ relaxor: dielectric spectroscopy evidence", *J. Phys.:Condens. Matter.* 19 (2007) 092001.

[50] P.W. Rehrig, S.E. Park, S.T. McKinstry, G.L. Messing, B. Jones, T.M. Shrout, "Piezoelectric properties of zirconium-doped barium titanate single crystals grown by templated grain growth", *J. Appl. Phys.*, 86 (1999) 1657-1661.

[51] Y. Zhi, R. Guo, A.S. Bhalla, "Growth of Ba(Ti$_{1-x}$Zr$_x$)O$_3$ single crystals by the laser heated pedestal growth technique", *J. Cryst. Growth*, 233 (2001) 460-465.

[52] Y. Zhi, R. Guo, A.S. Bhalla, "Dielectric behavior of Ba(Ti$_{1-x}$Zr$_x$)O$_3$ single crystals" *J. Appl. Phys.*, 88 (2000) 410-415.

[53] Y. Zhi, R. Guo, A.S. Bhalla, "Orientation dependence of the ferroelectric and piezoelectric behavior of Ba(Ti$_{1-x}$Zr$_x$)O$_3$ single crystals", *Appl. Phys. Lett.*, 77 (2000) 1535-1537.

[54] S. Wada, S. Suzuki, T. Noma, M. Kakihana, S.E. Park, L.E. Cross, T.R Shrout, "Enhanced Piezoelectric Property of Barium Titanate Single Crystals with Engineered Domain Configurations", *Jpn. J. Appl. Phys.*, Part 1, 38 (1999) 5505-5511.

[55] L.G.A. Marques, L.S. Cavalcante, A.Z. Simões, F.M. Pontes, L.S. Santos-Junior, M.R.M.C. Santos, I.L.V. Rosa, J.A. Varela, E. Longo, "Temperature dependence of dielectric properties for Ba(Zr$_{0.25}$Ti$_{0.75}$)O$_3$ thin films obtained from the soft chemical method", *Mater. Chem. Phys.*, 105 (2007) 293-297.

[56] D. Wu, Ph. Sciau, S. Schamm, F. Gloux, M. Fernandez, J.A. Varela, "Preparation and microstructures of BaTi$_{1-x}$Zr$_x$O$_3$ hetero-epitaxial thin films on SrTiO$_3$ substrates", *J. Phys. D: Appl. Phys.*, 40 (2007) 4701-4706.

[57] F.M. Pontes, M. T. Escote, C.C. Escudeiro, E.R. Leite, E. Longo, A.J. Chiquito, P.S. Pizani, A.J. Varela, "Characterization of BaTi$_{1-x}$Zr$_x$O$_3$ thin films obtained by a soft chemical spin-coating technique", *J. Appl. Phys.*, 96 (2004) 4386-4391.

[58] V. Reymond, "Structural and electrical properties of BaTi$_{1?x}$Zr$_x$O$_3$ sputtered thin films: effect of the sputtering conditions", *Thin Solid Films*, 467 (2004) 54-58.

[59] V. Reymond; S. Payan, D. Michau, J.-P. Manaud; M. Maglione, "Structural and electrical properties of BaTi$_{1-x}$Zr$_x$O$_3$ sputtered thin films: effect of the sputtering conditions", *Thin solid films*, 467 (2004) 54-58.

[60] T. M. Shaw, S. Trolier-McKinstry, P. C. McIntyre, "The Properties of Ferroelectric Films at Small Dimensions", *Annu. Rev. Mater. Sci.*, 30 (2000) 263-298.

[61] A. Dixit, S. B. Majumder, P. S. Dobal, R. S. Katiyard, A. S. Bhalla, "Phase transition studies of sol-gel deposited barium zirconate titanate thin films", *Thin Solid Films*, 447-448 (2004) 284-288.

[62] J. Bera, S.K. Rout, "On the formation mechanism of $BaTiO_3$–$BaZrO_3$ solid solution through solid-oxide reaction", *Mater. Lett.*, 59 (2005), 135-138.

[63] S. Gopalan and A.V. Virkar, "Interdiffusion and Kirkendall Effect in Doped $BaTiO_3$–$BaZrO_3$ Perovskites: Effect of Vacancy Supersaturation", *J. Am. Cerm. Soc.*, 82 (1999) 2887-2899.

[64] U. Weber, G. Greuel, U. Boettger, S. Weber, D. Hennings, R. Waser, „Dielectric Properties of $Ba(Zr,Ti)O_3$-Based Ferroelectrics for Capacitor Applications", *J. Am. Ceram. Soc.*, 84 (2001) 759-766.

[65] C.E. Ciomaga, M.T. Buscaglia, M. Viviani, V. Buscaglia, L. Mitoşeriu, A. Stancu, P. Nanni, "Preparation and dielectric properties of $BaZr_{0.1}Ti_{0.9}O_3$ ceramics with different grain sizes", *Phase Transit.*, 79 (2006) 389-397.

[66] S.M. Neirman, "The Curie point temperature of $Ba(Ti_{1-x}Zr_x)O_3$ solid solutions", *J. Mater. Sci.*, 23 (1988) 3973-3980.

[67] X.G. Tang, X.X. Wang, K.-H. Chew, H.L.W. Chan, "Relaxor behavior of $(Ba,Sr)(Zr,Ti)O_3$ ferroelectric ceramics", *Solid State Commun.*, 136 (2005) 89-93.

[68] X.G. Tang, J. Wang, X.X. Wang, H.L.W. Chan, "Effects of grain size on the dielectric properties and tunabilities of sol–gel derived $Ba(Zr_{0.2}Ti_{0.8})O_3$ ceramics", *Solid State Commun.*, 131 (2004) 163-168.

[69] M. Veith, S. Mathur, N. Lecerf, V. Huch, T. Decker, H.P. Beck, W. Eiser, R. Haberkorn, "Sol-Gel Synthesis of Nano-Scaled $BaTiO_3$, $BaZrO_3$ and $BaTi_{0.5}Zr_{0.5}O_3$ Oxides via Single-Source Alkoxide Precursors and Semi-Alkoxide Routes", *J. Sol-Gel Sci. Technol.*, 17 (2000) 145-158.

[70] S,K. Rout, T. Badapanda E. Sinha, S. Panigrahi, P.K. Barhai, P. T.P. Sinha, "Dielectric and phase transition of $BaTi_{0.6}Zr_{0.4}O_3$ ceramics prepared by a soft chemical route", *Appl. Phys. A*, 91 (2008) 101-106, 2008.

[71] H. Xu, L. Gao, H. Zhou, J. Guo, "Synthesis of nanopowders of $Ba(Ti_{0.8}Zr_{0.2})O_3$", *Mater. Lett.*, 58 (2004) 1999-2001.

[72] X. Cheng, H. Liu, Z. Li, Z. Yao, Y. Liu, Z. Yu, M. Cao, "Ceramic Processing Research Synthesis and characterization of $Ba(Ti_{1-x}Zr_x)O_3$ nanopowders by an aqueous coprecipitation method", *J. Ceram. Process. Res.*, 9 (2008) 576-579.

[73] S. Bhaskar Reddy, K. Prasad Rao, M.S. Ramachandra Rao "Nanocrystalline barium zirconate titanate synthesized at low temperature by an aqueous co-precipitation technique", *Scripta Materialia*, 57 (2007) 591-594.

[74] M.S. Dash, J. Bera, S. Ghosh, "Study on phase formation and sintering kinetics of $BaTi_{0.6}Zr_{0.4}O_3$ powder synthesized through modified chemical route", *J. Alloy Compd,*, 430 (2007) 212-216.

[75] A. Outzourhit, M.A. El Idrissi Raghni, M.L. Hafid, F. Bensamka, Abdelkader Outzourhit, "Characterization of hydrothermally prepared $BaTi_{1-x}Zr_xO_3$", *J. Alloy Compd.*, 340, (2002) 214-219.

[76] B.W. Lee, S.-B. Cho, "Preparation of $BaZr_xTi_{1-x}O_3$ by the hydrothermal process from peroxo-precursors", *J. Eur. Ceram. Soc.* 25 (2005) 2009-2012.

[77] A. Ianculescu, S. Guillemet-Fritsch, B. Durand, A. Brăileanu, M. Crişan, D. Berger, "$BaTiO_3$ nanopowders and nanocrystalline ceramics: I. Nanopowders", pp. 89-117, in *New developments in Advanced Functional Ceramics*, Ed. by L. Mitoşeriu, Transworld Research Network, India, 2007.

[78] A. Ianculescu, A. Brăileanu, M. Crişan, P. Budrugeac, N. Drăgan, G. Voicu, D. Crişan, V. Marinescu, „Influence of Barium Source on the Characteristics of Sol-Precipitated $BaTiO_3$ Powders and Related Ceramics", *J. Therm. Anal. Calorim.*, 88 (2007) 251-260.

[79] A. Ianculescu, A. Brăileanu, M. Crisan, L. Mitoşeriu, F. Tufescu, G. Voicu, N. Dragan, D. Crisan, E. Vasile, "Investigation of Microstructure and Properties of $BaTi_{0.85}Zr_{0.15}O_3$ Ceramics Prepared by a Sol-Precipitation Route", Proceedings 10th ECerS Conf., Göller Verlag, Ed. J.G.Heinrich and C. Aneziris, Baden-Baden, (2007) 575-580.

[80] A. Ianculescu, D. Berger, C. Matei, P. Budrugeac, L. Mitoşeriu, E. Vasile, "Synthesis of $BaTiO_3$ by soft chemistry routes", *J. Electroceram.*, 24 (2010) 46-50.

[81] M.P. Pechini, Method of preparing lead and alkaline earth titanates and niobates and coating method using the same for a capacitor. US Patent No. 3330697, 11 July 1967.

[82] D. Hennings, W. Mayr, "Thermal Decomposition of (BaTi) Citrate into Barium Titanate", *J. Solid State Chem.*, 26 (1978) 329-338.

[83] S. Kumar, G. L. Messing, "Synthesis of barium titanate by a basic pH Pechini process", *Mat. Res. Soc. Symp. Proc.*, 271 (1992) 95-100.

[84] S. Kumar, G.L. Messing, W.B. White, "Metal Organic Resin Derived Barium Titanate: I, Formation of Barium Titanium Oxycarbonate Intermediate", *J. Am. Ceram. Soc.*, 76 (1993) 617-624.

[85] S. Kumar, G. L. Messing, "Metal Organic Resin Derived Barium titanate: II. Kinetics of $BaTiO_3$ Formation", *J. Am. Ceram. Soc.*, 77 (1994) 2940-2948.

[86] J.P. Coutures, P. Odier, C. Proust, "Barium titanate formation by organic resins formed with mixed citrate", *J. Mat. Sci.*, 27 (1992) 1849-1856.

[87] M. Arima, M. Kakihana, M. Yashima, M. Yoshimura, "Polymerized complex of pure $BaTiO_3$ at reduced temperature", *Eur. J. Solid State Inorg. Chem., Soc.*, 32 (1995) 863-871.

[88] M. Arima, M. Kakihana, Y. Nakamura, M. Yashima, M. Yoshimura, "Polymerized Complex Route to Barium Titanate Powders Using Barium-Titanium Mixed-Metal Citric Acid Complex", *J. Am. Ceram. Soc.*, 79 (1996) 2847-2856.

[89] W.-S. Cho, "Structural Evolution and Characterization of $BaTiO_3$ Nanoparticles Synthesized from Polymeric Precursor", *J. Chem. Phys. Solids*, 59 (1998) 659-666.

[90] J.-D. Tsay, T.-T. Fang, T.A. Gubiotti, J.Y. Ying, "Evolution of the formation of barium titanate in the citrate process: the effect of the pH and the molar ratio of barium ion and citric acid", *J. Mater. Sci.*, 33 (1998) 3721-3727.

[91] P. Durán, F. Capel, J. Tartaj, C. Moure, "$BaTiO_3$ formation by thermal decomposition of (BaTi)-citrate polyester resin in air", *J. Mater. Res.*, 16 (2001) 197-209.

[92] P. Durán, F. Capel, D. Gutierrez, J. Tartaj, A. Baňares, C. Moure, "Metal citrate polymerized complex decomposition leading to the synthesis of $BaTiO_3$: effects of the precursor structure on the $BaTiO_3$ formation mechanism", *J. Mater. Chem.*, 11 (2001) 1828-1836.

[93] M.I.B. Bernardi, E. Antonelli, A.B. Lourenco; C.A.C. Feitosa, L. J. Q. Maia, A.C. Hernandes, "$BaTi_{1-x}Zr_xO_3$ nanopowders prepared by the modified pechini method", *J. Therm. Anal. Calorim.*, 87 (2007) 725-730.

[94] A. Ianculescu, L. Mitoşeriu, D. Berger, C.E. Ciomaga, D. Piazza, C. Galassi, "Composition dependent ferroelectric properties of $Ba_{1-x}Sr_xTiO_3$ ceramics", *Phase Transit.*, 79 (2006) 375–388.

[95] A. Ianculescu, D. Berger, M. Viviani, C. Ciomaga, L. Mitoşeriu, E. Vasile, N. Drăgan, D. Crişan, "Investigation of $Ba_{1-x}Sr_xTiO_3$ ceramics

prepared from powders synthesized by the modified Pechini route", *J. Eur. Ceram. Soc.*, 27 (2007) 3655-3658.

[96] A. Ianculescu, D. Berger, L. Mitoşeriu, L.P. Curecheriu, N. Drăgan, D. Crişan, E. Vasile, "Properties of $Ba_{1-x}Sr_xTiO_3$ ceramics prepared by the modified-Pechini method", *Ferroelectrics*, 369 (2008) 22-34.

[97] M. H. Frey, D. A. Payne, "Grain-size effect on structure and phase transformations for barium titanate", *Phys. Rev. B*, 54 (1996) 3158-3168.

[98] C.E. Ciomaga, "Contributions to the study of ferroelectric relaxors", PhD Thesis, "Al.I. Cuza" University, Iaşi, România (2007).

[99] J.T.S. Irvine, D.C. Sinclair, A.R. West, "Electroceramics: Characterisation by ac Impedance Spectroscopy", *Adv. Mater.* 2 (2004) 132-138.

[100] N. J. Kidner, Z. J. Homrighaus, B. J. Ingram, T. O. Mason, E. J. Garboczi, *"Impedance/Dielectric Spectroscopy of Electroceramics – Part 1: Evaluation of Composite Models for Polycrystalline Ceramics"*, J. Electroceram., 14 (2005) 283-291.

[101] A.R. James, C. Prakash, G. Prasad, "Structural properties and impedance spectroscopy of excimer laser ablated Zr substituted $BaTiO_3$ thin films", *J. Phys. D: Appl. Phys.* 39 (2006) 1635-164.

[102] G.A. Smolenskii, "Physical Phenomena in Ferroelectrics with Diffused Phase Transition", *Proc. Int. Meet. Ferroelectr.*, 2nd, 1969, (1970) 26-36.

[103] V.V. Kirilov, V.A. Isupov, "Relaxation Polarization of $PbMg_{1/3}Nb_{2/3}O_3$(PMN) – A *Ferroelectric* with A Diffused Phase Transition" *Ferroelectrics*, 5 (1973) 3-9.

[104] S.M. Pilgrim, A.E. Sutherland, S.R. Winzer, "Diffuseness as a Useful. Parameter for Relaxor Ceramics", *J. Am. Ceram. Soc.*, 73 (1990) 3122-3125.

[105] A.A. Bokov, Z.G. Ye, "Phenomenological description of dielectric permittivity peak in relaxor ferroelectrics", *Solid State Commun.*, 116 (2000) 105-108.

[106] K. Uchino, S. Nomura, *"Critical exponents* of *dielectric* constants in diffused phase transition crystals", *Ferroelectrics Lett.*, 44 (1982) 55-61.

[107] I.A. Santos, J.A. Eiras, "Phenomenological description of the diffuse phase transition in ferroelectrics", *J. Phys.: Condens. Matter*, 13 (2001) 11733-11740.

[108] L. Mitoşeriu, "Investigation of the role of chemical inhomogeneity and grain size effects towards the relaxor behaviour in perovskite systems", Ph. D. Thesis, Univ. of Genoa, Italy (2005).

[109] L. Mitoşeriu, C. Harnagea, P. Nanni, A. Testino, M.T. Buscaglia, V. Buscaglia, M. Viviani, Z. Zhao, M. Nygren, "Local switching properties of dense nanocrystalline BaTiO$_3$ ceramics", *Applied Phys. Lett.*, 84 (2004) 2418-2420.

[110] Z. Zhao, V. Buscaglia, M. Viviani, M.T. Buscaglia, L. Mitoşeriu, A. Testino, M. Nygren, M. Johnsson, P. Nanni, "Grain-size effects on the ferroelectric behavior of dense nanocrystalline BaTiO$_3$ ceramics", *Phys. Rev. B*, 70 (2004) 024107.

[111] M.T. Buscaglia, M. Viviani, V. Buscaglia, L. Mitoşeriu, A. Testino, P. Nanni, Z. Zhao, M. Nygren, C. Harnagea, D. Piazza, C. Galassi, "High dielectric constant and frozen macroscopic polarization in dense nanocrystalline BaTiO$_3$ ceramics", *Phys. Rev. B.*, 73 (2006) 064114.

[112] E.J. Abram, D.C. Sinclair and A.R. West, "A Strategy for Analysis and Modelling of Impedance Spectroscopy Data of Electroceramics: Doped Lanthanum Gallate", *J. Electroceram.*, 10 (2003) 165-177.

[113] D.P. Almond, A.R. West, "Ion Concentrations in *Solid* Electrolytes from an Analysis of AC Conductivity", *Solid State Ionics*, 9/10 (1983) 277-282.

[114] D.C. Sinclair, A.R. West, "Impedance and modulus spectroscopy of semiconducting BaTiO$_3$ showing positive temperature coefficient of resistance", *J. Appl. Phys.* 66 (1989) 3850-3856.

[115] D.C. Sinclair, A.R. West, "Electrical properties of a LiTaO$_3$ single crystal", *Phys. Rev. B*, 39 (1989) 13486-13492.

[116] P. Debye, "Polar Molecules", New York: Chemical Catalogue Company (1929).

[117] S. Cole, R. Cole, "Dispersion and absorption in dielectrics. I. Alternating current characteristics", *J. Chem. Phys.*, 9 (1941) 341-351.

[118] A.K. Jonscher, *Dielectric Relaxation in Solids*, London: Chelsea Dielectric Press (1983).

[119] H. Ihrig, D. Hennings, "Electrical transport properties of *n*-type BaTiO$_3$", *Phys. Rev. B*, 17 (1978) 4593-4599.

[120] M. Maglione, M. Belkaoumi, "Electron–relaxation-mode interaction in BaTiO$_3$:Nb", *Phys. Rev. B*, 45 (1992) 2029-2034.

[121] O. Bidault, P. Goux, M. Kchikech, M. Belkaoumi, M. Maglione, "Space-charge relaxation in perovskites", *Phys. Rev. B*, 49 (1994) 7868-7873.

[122] B.S. Kang, S.K. Choi, C.H. Park, "Diffuse dielectric anomaly in perovskite-type ferroelectric oxides in the temperature range of 400–700 °C", *J. Appl. Phys.*, 94 (2003) 1904-1911.

[123] C.E. Ciomaga *et al.*, unpublished work.

[124] F. Preisach, „Über die magnetische Nachwirkung", *Z. Phys.*, 94 (1935) 277-302.

[125] H.G. Katzgraber, F. Pazmandi, C.R. Pike, K. Liu, R.T. Scalettar, K.L. Verosub, G.T. Zimanyi, "Reversal-Field Memory in the Hysteresis of Spin Glasses", *Phys. Rev. Lett.*, 89 (2002) 257202.

[126] C. Enachescu, R. Tanasa, A. Stancu, E. Codjovi, J. Linares, F. Varret, "FORC method applied to the thermal hysteresis of spin transition solids: first approach of static and kinetic properties", *Physica B*, 343 (2004) 15-19.

[127] A. Stancu, D. Ricinschi, L. Mitoşeriu, P. Postolache, M. Okuyama, "First-order reversal curves diagrams for the characterization of ferroelectric switching", *Appl. Phys. Lett.*, 83 (2003) 3767-3769.

[128] D. Ricinschi, L. Mitoşeriu, A. Stancu, P. Postolache, M. Okuyama, "Analysis of the Switching Characteristics of PZT Films by First Order Reversal Curve Diagrams", *Integr. Ferroelectr.*, 67 (2004) 103-115.

[129] L. Cima, E. Labouré, "Characterisation and model of ferroelectrics based on experimental Preisach density", *Rev. Sci. Instrum.*, 73 (2002) 3548-3552.

[130] L. Cima, E. Labouré, "A model of ferroelectric behavior based on a complete switching density", *J. Appl. Phys.*, 95 (2004) 2659.

[131] A. Stancu, C. Pike, L. Stoleriu, P. Postolache, D. Cimpoeşu, "Micromagnetic and Preisach analysis of the First Order Reversal Curves (FORC) diagram", *J. Appl. Phys.*, 93 (2003) 6620-6622.

[132] D. Bolten, U. Böttger, R. Waser, "Reversible and irreversible polarization processes in ferroelectric ceramics and thin films", *J. Appl. Phys.*, 93 (2003) 1735-1742.

[133] D. Bolten, U. Böttger, R. Waser, "Reversible and irreversible piezoelectric and ferroelectric response in ferroelectric ceramics and thin films", *J. Eur. Ceram. Soc.*, 24 (2004) 725-732.

[134] C.H. Tsang, F.G. Shin, "Simulation of nonlinear dielectric properties of polyvinylidene fluoride based on the Preisach model", *J. Appl. Phys.*, 93 (2003) 2861-2865.

[135] Y. Yamashita, H. Yamamoto, Y. Sakabe, "Dielectric Properties of $BaTiO_3$ Thin Films Derived from Clear Emulsion of Well-Dispersed Nanosized $BaTiO_3$ Particles", *Jpn. J. Appl. Phys., Part 1*, 43 (2004) 6521-6524.

[136] A. Li, D. Wu, H. Ling, "Effects of processing on the characteristics of $SrBi_2Ta_2O_9$ films prepared by metalorganic decomposition", *J. Appl. Phys.*, 88 (2000) 1035-1041.

[137] H.-Y. Chou, T.-M. Chen, T.-Y. Tseng, "Electrical and dielectric properties of low-temperature crystallized $Sr_{0.8}Bi_{2.6}Ta_2O_{9+x}$ thin films on $Ir/SiO_2/Si$ substrates", *Mater. Chem. Phys.*, 82 (2003) 826-830.

[138] S.B. Ren, C.J. Lu, H.M. Shen, Y N. Wang, "In situ study of the evolution of domain structure in free-standing polycrystalline $PbTiO_3$ thin films under external stress", *Phys. Rev. B*, 55 (1997) 3485-3489.

[139] M. Nagata, D.P. Vijay, X.B. Zhang, S.B. Desu, "Formation and properties of $SrBi_2Ta_2O_9$ thin films", *Phys. Status Solidi*, A 157 (1996) 75-82.

[140] G. Arlt, D. Hennings, G. de With, "Dielectric Properties of Fine-Grained Barium Titanate Ceramics", *J. Appl. Phys.*, 58 (1985) 1619-1625.

[141] Y. Kinoshita, A.J. Yamaji, J. Appl. Phys., "Grain-size effects on dielectric properties in barium titanate ceramics", 47 (1976) 371-373.

[142] L. Mitoşeriu, C.E. Ciomaga, V. Buscaglia, L. Stoleriu, D. Piazza, C. Galassi, A. Stancu, P. Nanni, "Hysteresis and tunability characteristics of $Ba(Zr,Ti)O_3$ ceramics described by First Order Reversal Curves diagrams", *J. Eur. Ceram. Soc.*, 27 (2007) 3723-3726.

[143] D. Piazza, L. Stoleriu, L. Mitoşeriu, A. Stancu, C. Galassi, "Characterisation of porous PZT ceramics by first-order reversal curves (FORC) diagrams", *J. Eur. Ceram. Soc.*, 26 (2006) 2959-2962.

[144] L. Mitoşeriu, L. Stoleriu, M. Viviani, D. Piazza, M.T. Bucaglia, R. Calderone, V. Buscaglia, A. Stancu, P. Nanni, C. Galassi, "Influence of stoichiometry on the dielectric and ferroelectric properties of the tunable $(Ba,Sr)TiO_3$ ceramics investigated by First Order Reversal Curves method", *J. Eur. Ceram. Soc.* 26 (2006) 2915-2921.

[145] L. Stoleriu, A. Stancu, L. Mitoşeriu, D. Piazza, C. Galassi, "Analysis of switching properties of porous ferroelectric ceramics by means of first-order reversal curve diagrams", *Phys. Rev. B*, 74 (2006) 174107.

[146] D. Kolar, M. Trontelj, Z. Stadler, "Influence of interdiffusion on solid solution formation and sintering in the system $BaTiO_3$–$SrTiO_3$". *J. Am. Ceram. Soc.*, 65 (1982) 470-474.

[147] M.H. Frey, Z. Xu, P. Han, D.A Payne, "The role of interfaces on an apparent grain size effect on the dielectric Properties for *ferroelectric* barium titanate ceramics", *Ferroelectrics*, 206-207 (1998) 337-353.

[148] A.V. Polotai, A.V. Ragulya, C.A. Randall, "Preparation and Size Effect in Pure Nanocrystalline Barium Titanate Ceramics", *Ferroelectrics*, 288 (2003) 93-102.

[149] Z. Shen, Z. Zhao, H. Peng, M. Nygren, "Formation of tough interlocking microstructures in silicon nitride ceramics by dynamic ripening", *Nature*, 417 (2002) 266-269.

[150] B. Li, X. Wang, M.Cai, L. Hao, L. Li, "Densification of uniformly small-grained $BaTiO_3$ using spark-plasma-sintering", *Mater. Chem. Phys.*, 82 (2003) 173-180.

[151] B. Li, X. Wang, L. Li, H. Zhou, X. Liu, X. Han, Y. Zhang, X. Qi, X. Deng, "Dielectric properties of fine-grained $BaTiO_3$ prepared by spark-plasma-sintering", *Mater. Chem. Phys.*, 83 (2004) 23-28.

[152] R. Licheri, S. Fadda, Roberto Orrù, G. Cao, V. Buscaglia, "Self-propagating high-temperature synthesis of barium titanate and subsequent densification by spark plasma sintering (SPS)", *J. Eur. Ceram. Soc.*, 27 (2007) 2245-2253.

[153] S. Mustofa, T. Araki, T. Furusawa, M. Nishida, T. Hino, "The PLD of $BaTiO_3$ target produced by SPS and its electrical properties for MLCC application", *Mater. Sci. Eng.*, B103 (2003) 128-134.

[154] W. Luan, L. Gao, H. Kawaoka, T. Sekino, K. Niihara, "Fabrication and characteristics of fine-grained $BaTiO_3$ ceramics by spark plasma sintering", *Ceram. Int.*, 30 (2004) 405-410.

[155] T. Takeuchi, M. Tabuchi, H. Kageyama, "Preparation of dense $BaTiO_3$ ceramics with submicrometer grains by spark plasma sintering". *J. Am. Ceram. Soc.*, 82 (1999) 939-943.

[156] T. Takeuchi, C. Capiglia, N. Balakrishnan, Y. Takeda, H. Kageyama, "Preparation of fine grained $BaTiO_3$" ceramics by spark plasma sintering. *J. Mater. Res.*, 17 (2002) 575-581.

[157] X. Deng, X. Wang, H. Wen, A. Kang, Z. Gui, L. Li, "Phase transitions in nanocrystalline barium titanate ceramics prepared by spark plasma sintering", *J. Am. Ceram. Soc.*, 89 (2006), 1059-1064.

[158] M. Omori, "Sintering, consolidation, reaction and crystal growth by spark plasma sintering (SPS)", *Mater. Sci. Eng. A*, 287 (2000) 183-188.

[159] M. Nygren, Z. Shen, "On the preparation of bio-, nano- and structural ceramics and composites by spark plasma sintering", *Solid State Sci.*, 5 (2003) 125-131.

[160] M.T. Buscaglia, V. Buscaglia, M. Viviani, J. Petzelt, M. Savinov, L. Mitoşeriu, A. Testino, P. Nanni, C. Harnagea, Z. Zhao, M. Nygren,

"Ferroelectric properties of dense nanocrystalline $BaTiO_3$ ceramics", *Nanotechnology*, 15 (2004) 1113-1117.

[161] T. Hungría, M. Algueró, A.B. Hungría, A. Castro, "Dense fine-grained $Ba_{1-x}Sr_xTiO_3$ ceramic prepared by the combination of mechanosynthesized nanopowders and spark plasma sintering. *Chem. Mater.*, 17 (2005) 6205-6212.

[162] R. Waser, R. Hagenbeck, *Acta Mater.*, "Grain boundaries in dielectric and mixed-conducting ceramics", 48 (2000) 797-825.

[163] J. Xu, M. Itoh, *Chem. Mater.*, "Unusual dielectric relaxation in lightly doped n-type rhombohedral $BaTi_{0.85}Zr_{0.15}O_3$:Ta ferroelectric ceramics", 17 (2005) 1711-1716.

[164] G.A. Samara, L.A. Boatner, "Ferroelectric-to-relaxor crossover and oxygen vacancy hopping in the compositionally disordered perovskites $KTa_{1-x}Nb_xO_3$:Ca", *Phys. Rev. B*, 61 (2000) 3889-3896.

[165] 164. O. Bidault, M. Maglione, M. Actis, M. Kchikech, B. Salce, "Polaronic relaxation in perovskites", *Phys. Rev. B*, 52 (1995) 4191-4197.

[166] C. Ang, Y. Zhi, L.E. Cross, "Oxygen-vacancy-related low-frequency dielectric relaxation and electrical conduction in $Bi:SrTiO_3$", *Phys. Rev. B*, 62 (2000) 228-236.

[167] A.K. Jonscher, *Universal Relaxation Law* (Chelsea, London: Dielectric Press), 1996.

[168] I. Bunget, M. Popescu, *Physics of Solid Dielectrics*, Elsevier: New York, 1978.

[169] V. Petkov, V. Buscaglia, M. T. Buscaglia, Z. Zhao, Y. Ren, "Structural coherence and ferroelectricity decay in submicron- and nano-sized perovskites", *Phys. Rev. B*, 78 (2008) 054107.

[170] A. Yu. Emelyanov, N.A. Pertsev, S. Hoffmann-Eifert, U. Böttger, R. Waser, "Grain-boundary effect on the Curie–Weiss law of ferroelectric ceramics and polycrystalline thin films: calculation by the method of effective medium", *J. Electroceram.*, 9 (2002) 5-16.